Particle Searches and Discoveries—1976
(Vanderbilt Conference)

AIP Conference Proceedings
Series Editor: Hugh C. Wolfe
Number 30
Particles and Fields Subseries No. 11

Particle Searches and Discoveries—1976
(Vanderbilt Conference)

Editor
R.S. Panvini
Vanderbilt University

American Institute of Physics
New York   1976

ADVISORS AND SESSION CHAIRMEN

C. Baltay, Chairman, Neutrino Experiments

E. L. Berger

J. D. Bjorken

W. Chinowsky, Chairman, e^+e^- Experiments

J. Cronin, Chairman, Prompt Lepton Experiments

T. Ferbel, Chairman, Particle Searches in Hadroproduction and Photoproduction

C. Quigg, Chairman, Theory Session

B. Richter

J. Rosner

LOCAL COMMITTEE

R. S. Panvini

J. S. Poucher

S. L. Stone

A. H. Rogers

Copyright © 1976 American Institute of Physics, Inc.

This book, or parts thereof, may not be reproduced in any form without permission.

L.C. Catalog Card No. 76-19949
ISBN 0-88318-129-0
ERDA CONF- 760344

American Institute of Physics
335 East 45th Street
New York, N.Y. 10017

Printed in the United States of America

FOREWORD

This was the second international conference in high energy physics held at Vanderbilt University. The first was held in March, 1973, and dealt largely with new results from Fermilab and the CERN ISR. Experimental results on new particle searches and properties of the Psi/J family of particles was the central theme of the present conference. The topics of the experimental sessions were: searches for new particles in hadro- and photoproduction, prompt lepton data, neutrino experiments, and e^+e^- experiments.

The total attendance was just over one hundred and invitations were open to anyone who wished to attend. The conference was organized by soliciting advice from a group of active researchers who were asked to comment on the content of the conference and to suggest other people to contact and/or give talks. The session chairmen, chosen from among the most active people in their areas of expertise, were responsible for issuing further invitations and organizing the schedules of their sessions. Preparations at Vanderbilt, including the banquet, lecture hall arrangements, registration procedures, etc., were organized by a local committee of Vanderbilt faculty. The list of names of people who contributed in all the above-mentioned areas is given on a previous page in these proceedings. Special thanks are due also to other members of the Vanderbilt community who assisted in making things go smoothly. In particular, we are indebted to S. Bailey and M. Duro, who ably managed matters at the registration desk, and to R. Hichwa and J. Patrick who made the question/answer periods more effective by their handling of the microphones. A large part of the social success of the conference is due to Virginia Holladay, who organized the tours of Nashville for people who accompanied participants, and to Sylvia Rogers, who planned our gourmet banquet.

Financial support for the 1976 Vanderbilt Conference was provided by the National Science Foundation, the United States Energy Research and Development Administration, and from special funds provided by the Vanderbilt University Research Council.

R. S. Panvini

May, 1976

CONTENTS

- ADVISORS, SESSION CHAIRMEN, LOCAL COMMITTEE
- FOREWORD
- CONFERENCE PROGRAM AND INDEX TO PAPERS

Monday, March 1, 1976

Morning Session

I. Welcoming Statement

    Dean J. Voegeli, College of Arts and Sciences   *

II. Photoproduction and Hadroproduction of New Particles

    Chairman: T. Ferbel, University of Rochester

1) Photoproduction and Hadroproduction of $\mu^+\mu^-$ pairs at Fermilab

    I. Gaines, Fermilab   *

2) Production of High Mass e^+e^- Pairs at Fermilab

    J. Appel, Fermilab   1

3) Search for New Particles Below 50 GeV

    M. Chen, M.I.T.   13

4) A Search for New Particles in Proton-Nucleus Collisions at 400 GeV/c

    E. Shibata, Purdue   30

5) A Search for New Particles at the CERN ISR

    N. McCubbin, Rutherford Lab   38

6) Charmed Particle Searches Using Bubble Chambers with Hadronic Beams

    R. Harris, Fermilab   51

7) Implications for New Particle Searches for Some Cosmic Ray Emulsion Events

    T. Gaisser, Bartol Institute/BNL   57

Afternoon Session

III. Theory - Chairman: C. Quigg, Fermilab

    1) M.I.T. Bag Model - Spectroscopy

       K. Johnson, M.I.T.   *

    2) Nonperturbative Methods in Field Theory

       M. Weinstein, SLAC   *

    3) Comments on the New Particles

       T. Appelquist, Yale   62

Evening Session

IV. Photoproduction and Hadroproduction of New Particles (continued)

Chairman: T. Ferbel, University of Rochester

    8) ψ and Excess Leptons in Photoproduction

       D. Ritson, SLAC   75

    9) Production of the New Particles Using Pion and Proton Beams

       G. Sanders, Princeton   *

   10) Search for New States Which Decay Semi-Leptonically

       H. Lubatti, University of Washington   83

   11) Study of Prompt Muons and Associated Particles in n-Be Collisions

       R. Ruchti, Northwestern   88

   12) Dimuon Production by Pions and Protons in Iron and a Search for the Production in Hydrogen of New Particles which Decay into Muons

       M. Mallary, Northeastern   94

   13) Search for Narrow High-Mass States at Fermilab

       J. Matthews, Michigan State University   102

       Comment on Session Concerning New-Particle Searches
       T. Ferbel, University of Rochester   107

Tuesday, March 2, 1976

Morning Session

I. Prompt Lepton Production

   Chairman: J. Cronin, University of Chicago

   1) Prompt Positrons at AGS Energies--Report on the Penn-Stony Brook Experiment

      J. Kirz, Stony Brook 109

   2) Direct Production of Muons in the Forward Direction by 300 GeV Protons on Uranium

      R. Johnson, Fermilab 116

   3) Review of CERN-Columbia-Rockefeller-Saclay Experiment

      S. Segler, Rockefeller University 123

   4) Dimuon Production and Single Muon Polarization

      L. Leipuner, BNL .... *

   5) Production of Direct Positrons at 256 and 800 MeV

      R. Mischke, Los Alamos 134

   6) Direct Muon and Electron Production at ISR

      C. Rubbia, Harvard .... *

   7) Review of Theoretical Ideas on Prompt Lepton Production

      J. D. Sullivan, University of Illinois 142

Afternoon Session

II. Neutrino Experiments

   Chairman: C. Baltay, Columbia University

   1) Neutrino Interactions in H_2 and D_2 in BNL 7 Foot Chamber

      E. Cazzoli, BNL 162

   2) Observation of the Reaction $\nu p \to \nu p$

      C. Pang, University of Illinois 173

3) The Neutral Current Couplings

   B. Barish, Caltech 180

4) Anomalies in ν, $\bar{\nu}$ Interactions

   A. Benvenuti, University of Wisconsin 189

5) New Phenomena in ν Interactions

   J. Mapp, University of Wisconsin 206

6) Search for New Particle Production in a $\bar{\nu}H_2$ Experiment and in a H_2-Ne Experiment in the 15-foot Bubble Chamber

   W. Scott, Fermilab 217

Wednesday, March 3, 1976

Morning Session

I. e^+e^- Experiments – ψ and ψ' Decays and Intermediate States

Chairman: W. Chinowsky, LBL

1) ψ and ψ' Decays and Intermediate States: Results from SPEAR, LBL-SLAC

   C. Friedberg, LBL 229

2) Results from DORIS, DASP

   K. Pretzl, Max Planck Institute 248

3) Results from DORIS, Neutral Detector

   H. Rieseberg, Heidelberg 274

II. e^+e^- Experiments – Other Results

4) Inclusive Particle Production and Anomalous Muons in e^+e^- Collisions at SPEAR

   B. Barnett, Johns Hopkins/Maryland 285

5) Other Structures in e^+e^- Annihilation

   M. Breidenbach, SLAC 305

Postlude

C. Quigg, Fermilab   321

LIST OF PARTICIPANTS   336

Papers marked with an asterisk instead of a page number were presented
at the conference but were not prepared for publication in the proceedings.

PRODUCTION OF HIGH MASS e^+e^- PAIRS AT FERMILAB

J. A. Appel
Columbia - Fermilab - Stony Brook Collaboration*

ABSTRACT

Electron-positron pairs with an effective mass up to 10 GeV have been observed in proton-beryllium collisions at 400 GeV at Fermilab. In addition to high mass continuum events, evidence is presented for a new resonance at 6 GeV. The J/ψ is observed and its production dynamics measured near x of 0 and out to P_T of 1.6 GeV. The decay angular distribution is found to be best fit with $(1 + \cos^2\theta^*)$. A 90% C.L. upper limit of 2% is placed on the branching ratio times cross section for ψ' relative to ψ.

INTRODUCTION

As the continuation of a long program of lepton pair measurements initiated by Leon Lederman at Brookhaven National Laboratory many years ago,[1] a Columbia-Fermilab group, with Professor Lederman as spokesman, has continued these measurements at Fermilab at higher energy and higher mass range. In hindsight, it appears that the previous measurements suffered from a lack of mass resolution, the price paid for high rate capability. Even before our hindsight received the focusing power of the J/ψ,[2] our new proposal at Fermilab turned to detecting electrons rather than muons and concentrated its efforts on mass resolution. This was in the hope of increased sensitivity for structure in the transverse momentum (p_T) spectrum of single electrons, a signature of the two body decay of such heavy objects as charged and neutral intermediate vector bosons, heavy leptons, and Lee-Wick bosons. From the beginning, the electron-positron pair measurements were part of the proposal, both as a measure of the significance of the single lepton signal observed and as a further probe of hadronic structure.

As is now well known, no structure has been observed in the spectrum of single leptons. With the possible exception of low p_T, an amazingly constant 10^{-4} ratio of direct leptons to pions independent of S and p_T has been

*J. A. Appel, B. C. Brown, C. N. Brown, D. C. Hom, W. R. Innes, D. M. Kaplan, L. M. Lederman, H. P. Paar, H. D. Snyder, J. M. Weiss, T. Yamanouchi, J. K. Yoh. Work supported by U.S.E.R.D.A. and N.S.F.

observed.[3] Rather than concentrate on the model-dependent limits of new particle production determined from single leptons, I will focus on the direct results of the electron-positron pair measurements. We have found these new results exciting both for the general feature of observing high mass events to 10 GeV and for the suggestion of a new resonance at 6 GeV which has been called upsilon.

APPARATUS

The apparatus used (Figure 1) to observe electron-positron pairs is an enlargement of the same ideas which were used in our single arm measurements[4]: combined lead glass and magnetic spectrometers, separation of magnetic deflection and production angles into vertical and horizontal planes, large momentum acceptance in both arms, and an increased rate capability by taking measurements only after the magnets, where the high energy particles of interest have been separated from the very intense neutral and lower energy charged fluxes. The observed particles are viewed through a 7 mr vertical by 45 mr horizontal aperture on each arm. The acceptance of each arm includes the region around 90° in the center of mass. The apparatus is symmetric except that opposite sign particles are bent up and down, respectively, on opposite sides of the continuation of the incident beam line. Standard trigger counters and vertical scintillation hodoscope elements are located at U1-3 and D1-3 as well as horizontal and crossed small angle stereo proportional wire chambers. These are used to re-

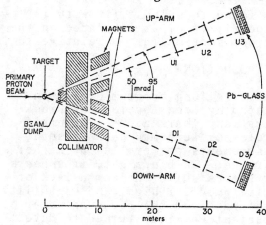

Figure 1. Schematic Diagram of Apparatus.

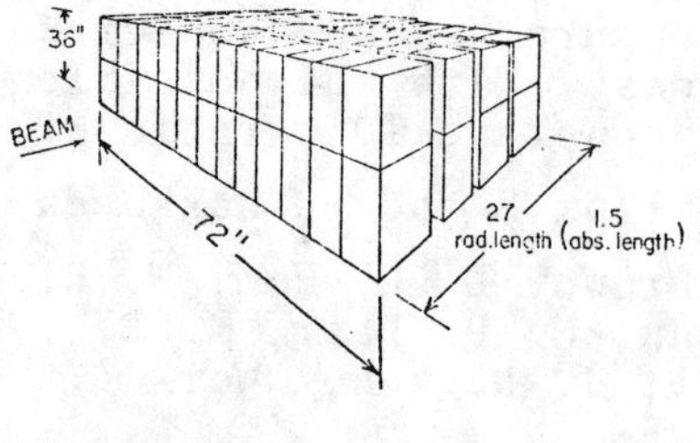

Figure 2. Lead Glass Array.

construct tracks downstream of the magnet and, using the target origin of the charged particles, calculate their momentum.

Electron and positron identification is accomplished using lead glass calorimetry. Ninety six blocks of lead glass, each with its own photomultiplier and signal processor, are arranged in each arm as shown in Figure 2. The electromagnetic shower associated with each entering electron is completely contained (to better than 99%) in the 27 radiation lengths of lead glass and 2 radiation lengths of lead placed in front of the array. Hadrons typically deposit only a small fraction of their energy in the 1.5 absorption lengths of lead glass and are easily rejected as electron candidates. Even those hadrons which deposit all their energy in the array can be identified by the distribution of energy deposition since the signal from each block is recorded. The hadron rejection in this experiment is better than 1/2500 on each arm. Figure 3 shows the spectrum of charged particles entering the glass which pass the kinds of shower development criteria I have just described. The distribution is shown as a function of the fraction of its energy that each particle deposits in the array. A clear electron peak is visible. If a beam devoid of electrons is sent into the array,[5] the distribution is like that of the dotted line which indicates that the electron signature is produced by true electrons and not the cuts applied to the data. We generally turn to this E/p distribution to learn the reliability of the electron identification. The final cut is then made on the ratio E/p for each particle.

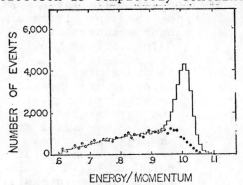

Figure 3. Spectrum of particles which pass shower cuts.

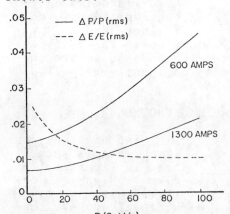

Figure 4. Apparatus Resolution

In addition to the particle identification properties of the lead glass system, its inherent energy measuring properties for electrons can be used to improve the momentum resolution obtained by the proportional wire system. As indicated by the resolution curves of Figure 4, the lead glass energy resolution improves at the higher momenta, where the PWC system is limited by the effects of finite wire spacing.

DATA ACCUMULATION

The large momentum acceptance in each arm of the spectrometer results in the characteristically large mass acceptances shown in Figure 5. Since the sensitivity of each measurement is determined by the lowest mass accepted in the apertures, runs have been taken with several settings of the current in the magnets. The data I am showing here comes from the running indicated in Table I. In general, the intensity was adjusted so that the charged particle coincidence rate was held fixed. When the beam spill improved, we increased the primary intensity correspondingly and vice versa. The data on the ψ was taken only with the 600 and 800 ampere running and the high mass data comes dominantly from 1100 and 1300 ampere running.

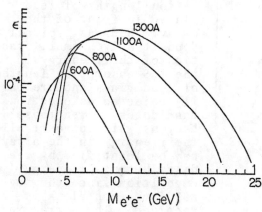

Figure 5. Mass Acceptance

SIGNAL AND UNPHYSICAL BACKGROUNDS

Table I

RUNNING CONDITIONS

CURRENT	600A	800A	1100A	1300A
BEAM INT. Protons/Pulse	2×10^9	6×10^9	2×10^{10}	4×10^{10}
INT. IN EACH ARM /Pulse	1.5×10^6	1.5×10^6	1.5×10^6	1.5×10^6
CHARGED PART. COIN. RATE /Pulse	5×10^4	5×10^4	5×10^4	5×10^4
TOTAL INTERACTIONS IN TARGET	6.5×10^{12}	2.2×10^{13}	1.3×10^{14}	2.3×10^{14}

The electron-positron data is summarized in Figure 6.[6] A clear ψ signal is observed with an rms width of 40 MeV, consistent with our estimated apparatus resolution. The dashed curve drawn under the histogram is our estimate of the nonphysical background for the data; i.e., the combination of accidental coincidences of particles

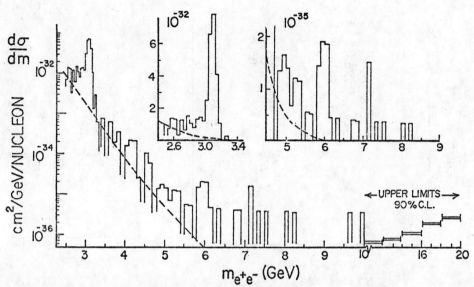

Figure 6. Electron-positron pair mass spectrum.

from separate interactions in the target and of true coincidences from a single interaction in the target, but where one or both of the observed particles was the result of a misidentified hadron, or of the conversion of a photon in the material upstream of the magnets. In the region of the ψ, the background is dominated by accidental coincidences and, at the level of 5%, there is no evidence of the physically interesting continuum. Taking five events as the upper limit on the $\psi'(3700)$, we obtain a 90% C.L. limit on the total cross section times branching ratio of $\sigma_T \cdot B(3700) / \sigma_T \cdot B(3100) \leq 2\%$. Above this region, the unphysical background (i.e., not directly produced electron-positron pairs) comes to be dominated by correlated processes in which multihadron states lead to two particles which masquerade as or turn into electrons and positrons of high effective mass.

In this experiment we have genuine monitors of these backgrounds from the actual data recorded on tape at the same time as the true electron-positron signal is obtained. The electron identification criteria are very loose in the trigger which leads to recording data on magnetic tape. This accounts for a large number of hadrons which appear in E/p plots below the 1.0 region. Some of these considerations are best indicated by Figure 7a which shows a scatter plot of some of the 600 ampere data. A clear ψ clustering is apparent where both arms give good electron identification. The size of the misidentified hadron pair background is also easy to estimate from this graph.

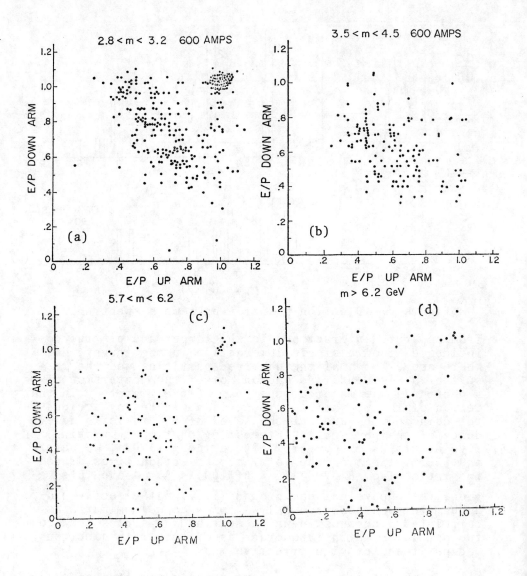

Figure 7. E/p Scatter plots for events in various mass ranges.

All one needs is the shape of the hadron distribution as a function of E/p (as given by the dotted line in Figure 3). By (1) taking the events for which both arms have signatures in the lower E/p regions for normalization and (2) estimating the number of events under the electron signal at E/p near 1.0, we get directly the accidental and correlated misidentified charged hadron pair background. Single arm triggers are taken at the same time as the pair data. These data are then combined off line to get the shape of the accidental coincidence background. The normalization is obtained by using the pair and single arm charged particle rates which I refered to earlier. Finally, in separate runs, thin converters are inserted into the spectrometer arms 6 meters downstream of the target. This results in a four-fold increase of the electron rates in each arm and a factor of 16 in the pairs due to photon conversions. These special runs give upper limits on the conversion electron pair contributions to the mass spectra and, using electron identification cuts as I have indicated, also give conversion electron-misidentified hadron pair backgrounds.

We have obtained two results from the study of the $\psi(3100)$: (1) new data on the dynamics of hadronic production of ψ's in previously unmeasured regions and on the ψ decay angular distribution and (2) confidence in our understanding of the acceptance of the apparatus. While the $\psi(3100)$ is well beyond the start of the 600 ampere acceptance, the start of the mass acceptance is near the ψ for the 800 ampere data. The x acceptance is near x of 0 and the p_T acceptance extends to p_T of 2 GeV. Using the data shown in Figure 8 we find the $\psi(3100)$ production cross section to be well represented by[7]

$$E\frac{d^3\sigma}{dp^3} \cdot B = (1.7 \pm 0.4) \times 10^{-32} e^{-cp_T^2} \text{ cm}^2/\text{GeV}^2/\text{Be Nucleus}$$

with $c = (1.1 \pm 0.35)$ GeV^{-2} at $x \simeq 0$.   (1)

$$(\frac{d\sigma}{dx})_{x=0} \cdot B = (2.6 \pm 0.7) \times 10^{-31} \text{ cm}^2/\text{Be Nucleus}$$

for $-.06 < x < +.08$ (2)

Using $E\frac{d^3\sigma}{dp^3} \propto (1-|x|)^{4.3}$ and linear A-dependence

$$\sigma_T \cdot B = (1.1 \pm 0.3) \times 10^{-32} \text{ cm}^2/\text{Nucleon}$$ (3)

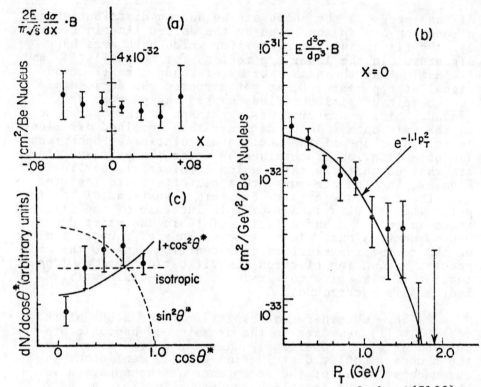

Figure 8. Production and Decay Dynamics of the ψ(3100).

Both the 600 and 800 ampere data are separately consistent with these values. A fit to the decay angular distribution $(1 + \cos^2\theta^*)$ has a 20% confidence level while an isotropic distribution cannot be ruled out (5% confidence level). A $\sin^2\theta^*$ distribution is ruled out.

With some confidence in the apparatus and having some physics under our belts we turn to the high mass region. Electron-positron pairs have been observed with masses as large as 10 GeV! There must be something hard in nucleons which allows such heavy objects to be produced. In addition there is the clustering of up to 12 events within our 100 MeV (rms) mass resolution in the region near 6 GeV, 12 out of the 27 events above 5.5 GeV!

First, let's see how convincing these events are as electron-positron pair events. Figure 7 shows the scatter plot of events in the region above the ψ(3100). In the region near 6 GeV and in the region above, clear electron-positron pair signatures are seen above the charged hadron pair background. No significant unphysical backgrounds occur in this mass region. Similarly, the test for other background sources proved negative as indicated by the dashed line in Figure 6. We have a clear electron-positron signal...whether a continuum alone or with resonant structure is another question.

Because of the large number of empty bins (whose size is indicative of the mass resolution), the standard statistical tests are not valid. Using bins much wider than the apparatus resolution allows us to make a number of distributions which are consistent with the wide binned data and with the total number of events. Using those distributions and the acceptance of the apparatus, events are thrown by Monte Carlo techniques and a search is made for the frequency of finding 11 events in three adjacent 100 MeV bins. From such distributions with mass dependances from $1/m^3$ to $1/m^7$ we get the conservative observation that the chances are less than 1/50 that the type of clustering we observe at 6 GeV is produced by a continuum consistent with the data observed between 5.5 and 10 GeV. Furthermore, some peaking at 6 GeV is evident in both 1100 and 1300 ampere data (Figure 9). One 6 GeV event even comes from the 600 ampere data! Thus, given a clear pair signal above the nonphysical background with a peaking which favors the interpretation as a new resonance evident in more than one magnet setting, we have proposed the name upsilon (ϒ) for this 6 GeV object.

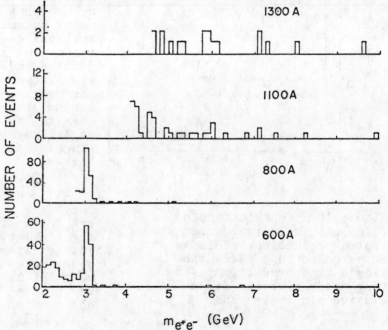

Figure 9. Event Distribution by Magnet Current

SIGNIFICANCE OF THE DATA

At this point, absolute normalizations are uncertain to 25% and we are putting a 50% uncertainty on our background estimates. Nevertheless, it is interesting to consider the relavance of these measurements to the single lepton production observed earlier and to various predictions of dilepton mass distributions from recent parton models. Given the fairly complete picture of ψ production and decay dynamics, it is appropriate to calculate the ψ contribution to single leptons. Figure 10 suggests that the ψ might account for a significant amount of the single leptons above p_T of 1.5 GeV. Below this value there would have to be other sources which blend more or less smoothly with the ψ contribution. Finally, the pair data above 5 GeV mass are comparable to the predictions of recent models, given the assumptions about color and with the anti-parton distributions as indicated in Figure 11. These high mass pairs (continuum and upsilon) seem to contribute negligibly to the single leptons at high p_T.

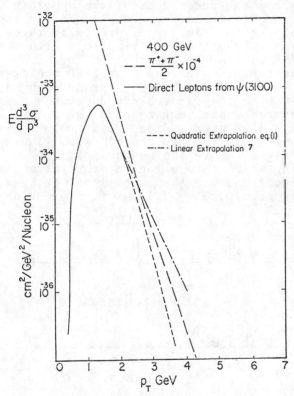

Figure 10. Contribution of ψ (3100) to inclusive direct lepton production. This is compared to 10^{-4} times the single pion production as a rough approximation of inclusive single leptons.

In conclusion, the present data are interesting as regards the possible existence of a new resonance, which we have dubbed upsilon, and the observation of continuum high mass electron-positron pairs up to 10 GeV in mass.

The apparatus has just been converted to observe muon pairs in the hopes of increasing the data rate - at some expense in mass resolution and event signature. We are very anxious to obtain more data in the upsilon region and at higher masses still.

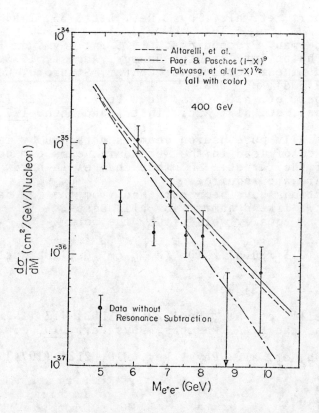

Figure 11. High mass pair data compared to theoretical models.[8]

NOTATION

B = branching ratio for e^+e^- decay

S = center of mass energy squared

p_T = momentum component transverse to incident beam

$p_{||}$ = momentum component paralled to incident beam

$x = p_{||}^{cm}/p_{max}^{cm}$

θ^* = decay angle in ψ helicity frame

FOOTNOTES AND REFERENCES

1. J. H. Christenson, et. al., Phys. Rev. Lett. <u>25</u>, 1523 (1970).
2. J. J. Aubert, et. al., Phys. Rev. Lett. <u>33</u>, 1404 (1974).
3. L. M. Lederman, Proceedings: 1975 Int. Sym. on Lepton and Photon Int. at High Energy, August, 1975. (Stanford Linear Accelerator Center, Stanford, California 94305), pg. 265.
4. J. A. Appel, et. al., Phys. Rev. Lett. <u>33</u>, 722 (1974). J. A. Appel, et. al., Nucl. Instr. and Meth. <u>127</u>, 495 (1975).
5. An essentially pure hadron beam is obtained by inserting 5.1 cm of lead in the secondary beams 6 m downstream of the target. No other changes in hardware or software are required.
6. Cross sections have been obtained from the raw data assuming ψ-like dynamics for all masses.
7. Our data is equally well fit by the form

$$E\frac{d^3\sigma}{dp^3} \cdot B = (2.5 \pm 0.6) \times 10^{-32} e^{-bp_T} \text{ cm}^2/\text{GeV}^2/\text{Be Nucleus}$$

with $b = (1.6 \pm .35)$ GeV^{-1}.

8. G. Altarelli, et. al., Nucl. Phys. <u>B92</u>, 413 (1975). H. P. Paar and E. A. Paschos, Phys Rev. <u>D10</u>, 1502 (1974). S. Pakvasa, et. al., Phys. Rev. <u>D10</u>, 2124 (1974).

SEARCH FOR NEW PARTICLES BELOW 50 BeV

Min Chen
Laboratory for Nuclear Science and Department of
Physics, Massachusetts Institute of Technology,
Cambridge, Massachusetts 02139

ABSTRACT

A systematic search for charm particles was made using the MIT-BNL double arm spectrometer, in the h^+h^-, $h^+h^-\mu$ and $h^{\pm}e^{\mp}$ final states with 28.5 GeV incident proton beam interacting on Be target. No evidence was found in any of the reactions.

INTRODUCTION

Should the newly discovered particles $J'(3.1)$[1,2,3], $\psi(3.7)$[4] etc. be explained as particles with some hidden quantum number, e.g. charm, then particles with that explicit quantum number should also exist. The MIT-BNL group has completed a systematic search for such particles in the following reaction:

$$P + Be \rightarrow h^+ + h^- + X$$
$$h^{\pm} + e^{\pm} + X$$
$$\mu^{\pm} + h^{\pm} + h^{\mp} + X$$

where $h^{\pm}$ stands for hadrons ($\pi^{\pm}$, $k^{\pm}$, $p^{\pm}$)

(I) In the h^+h^- mode

A systematic search in the following reactions in the pair mass region 1.2 - 5.5 BeV with a mass resolution of 5 MeV was made:[5]

$$p + p \rightarrow \pi^- p + X$$
$$\pi^+\pi^- + X$$
$$\bar{p}p + X$$
$$K^- p + X$$
$$K^+\pi^- + X$$
$$K^+K^- + X$$
$$K^-\pi^+ + X$$
$$K^+\bar{p} + X$$
$$\pi^+ p^- + X$$

The experiment set up (Fig.1,2) for this search is very similar to the original e^+e^- experiment where the J particles was first observed in August 1974. However, since there were many more hadrons than electrons the random accidentals were more serious. To reduce the accidentals to the minimum a new target system was put in. It consisted of 5 pieces of 4 mm x 4 mm x 4 mm Be target each separated by six inces. The targets were supported by thin piano wires. This arrangement enabled one to locate the point of intersection between two trajectories. Comparison with the target location enabled us to reduce random accidentals. To further reduce accidentals additional scintillation counters were installed to tighten the two arm coincidence to 0.9 ns. Two high pressure (300 psi) Cerenkov counters were installed to identify K's, replacing the shower counters. The counters C_e, C_o were filled with 1 atm isobutane to identify π's. The Cerenkov counters set the mass acceptance to 1 BeV. In this way all 9 combinations were measured simultaneously. To avoid systematic errors, 7 overlapping magnet settings were made for the measurement.

Figures 3 to 5 show the results of some of these typical reactions. Without acceptance corrections the yield increases with mass due to an increase in acceptance. It then decreases due to the decline in production cross sections. There are no sharp sharp narrow resonances in any of the 9 reactions. There may of course be very wide "ordinary" resonances with widths of 300 MeV or more. A search for these depends on exact calculations of acceptance and as yet have not been made.

To gain a feeling for the sensitivity of the measurements, we take the production mechanisms of the 9 reactions to be the same as that assumed for J. From this we obtain the following table:

Table I Sensitivity (cm^2) For Narrow Resonances (in unit for σ_J BR = 10^{-34} cm^2 at 28.5 GeV)

h^+x^- \ m	2.25	3.1	3.7 GeV
π^+K^-	10	.4	0.1
$K^+\pi^-$	40	.8	0.4
$p\bar{p}$	-	4	0.2
K^+K^-	10	0.5	0.1
$\pi^+\bar{p}$	20	.4	..07
$\pi^+\pi^-$	80	5	0.3
K^-p	70	4	.3
π^-p	400	40	5
$K^+\bar{p}$	20	.4	.08

Whereas the spectra do not show any sharp resonance states, the cross sections $\frac{d\sigma}{dm}$ vs m for groups A ($\pi^-\bar{p}$), B ($\pi^+\pi^-$, $\bar{p}p$, $K^+\pi^-$, K^-p) and C (K^+k^-, $K^+\bar{p}$, $K^-\pi^+$) do exhibit some simple degeneracies above the mass of J. The cross sections for each group decrease with a mass e^{-5m} and differ from each other by an order of magnitude (Fig. 6)[6]

The cross sections in the pair range from 3 to 5.5 GeV is shown in Fig. 7. This is the energy region where the ratio $e^+e^- \to$ hadron/$\mu^+\mu^-$ increases from 2.5 to 5. However the pair cross section measured here showed monotonically decreasing with the same slope e^{-5m} as in the low mass region.

(II) In the electron-hadron mode

With the right arm of the spectrometer tuned to detect hadrons, the left arm was converted to detect electrons (positrons): Both of the Cerenkov counters were filled with hydrongen gas and the lead glass and shower counters are put behind C_k to measure the pulse height of electromagnetic showers.

We scanned through the pair mass region from 1.5 to 3 GeV in three settings. The polarity of the spectrometer was reversed several times to measure both h^+e^- and h^-e^+ pairs.

Fig. 8 shows the relative timing of the electron and the hadron as measured in the spectrometer. We see a clear coincidence signal due to π^+e^- and π^-e^+. Some typical mass spectra of π^+e^- and π^-e^+ are shown in Fig. 9. A comparison of the genuine coincidence pairs with random accidental πe pair shows no difference in the mass spectra. In other words there is no indication of a discontinuity in the mass spectra due to some resonance decay such as $D \to e + h + x$.

Another interesting feature is the missing of the genuine coincidence of the signal due to k^-e and $\bar{p}e$ which were recorded at the same time as the π^-e events. In the hadron momentum range between 3.0 and 6.0 GeV the upper limit of k^-e^+ and $\bar{p}e^+$ are

$$\frac{\sigma(k^-e^+)}{\sigma(\pi^-e^+)} < 1\%$$

and

$$\frac{\sigma(\bar{p}e^+)}{\sigma(\pi^-e^+)} < 1\% \quad (95\% \text{ confidence level})$$

During the same running periods, 8 e^+e^- events with mass = 3.1 ± 0.007 were also observed. These events are clearly identified to be the J particle. Therefore the upper limit of the k^-e^+, $\bar{p}e^+$ channel can also be normalized to the production cross section of the J. We obtain (90% confidence level)

$$\frac{\int_2^3 \frac{d\sigma}{dm}(p+Be \to k^- + e^+ + x) \times (\text{Acceptance})\,dm}{\sigma_J(p+Be \to J+x) \times \text{Acceptance}_{e e}} < 7$$

$$\hookrightarrow e e$$

and

$$\frac{\int_{2.3}^{3.3} \frac{d\sigma}{dm}(p+Be \to \bar{p} + e^+ + x) \times \text{Acceptance}\,dm}{\sigma_J(p+Be \to J+x) \times \text{Acceptance}_{ee}} < 1$$

where m is in GeV.

(III) **In the $\mu h^+ h^-$ mode**

As we discussed in (I) that charm was not observed in the h^+h^- mode. There are however two possible explanations that charmed particles may be produced abundantly yet there is no significant signal in the h^+h^- mode: 1) The branching ratio of two body decay mode may be very low; 2) The noncharmed h^+h^- cross sections are much larger than the charmed ones such that the latter is swamped.

In order to enhance the signal due to charm, the following observations were made:

1) In hadron interactions charmed particles are produced in pairs in order to conserve quantum number.

2) There are theoretical speculations that the leptonic or semileptonic decay modes are of substantial size (8%).

Therefore an event selection for h^+h^- with a μ fragment from the other charm partner detected will be a much more selective signature of charm. One such possible reaction is

$$p + p \rightarrow D + \bar{D} + x$$
$$ \hookrightarrow k\pi \quad \text{(or } kp \text{ if Baryon)}$$
$$ \hookrightarrow k\,\mu\nu \quad \text{(or } \Lambda\mu\nu \text{ if Baryon)}$$

MUON DETECTOR

One of the unique features of the MIT-BNL double arm spectrometer is that it can withstand a very intense beam. Hence, the muon detector should have the following properties:

1) Operate under as intense hadron fluxes as the double arm spectrometer can.

2) Detect muons with CM momentum $p^* \sim \dfrac{M_D}{3} = 700$ MeV, where the mass of charmed mesons is expected to be $M_D \sim 2$ GeV.

3) Cover a large solid angle.

4) Absorb kaons and pions right next to the target, to reduce muon background from decays.

Figure 10 shows the plan and side view of detector. Tungsten blocks are piled around the beam 1" away from the target. A sealed one cubic meter box of uranium surrounds the tungsten collimator to further absorb the hadron shower. Beyond the uranium is one meter of lead concrete sandwich reducing the number of neutrons to a tolerable level. The detector covers a solid angle of 2 sr. and accepts muon momenta greater than 2.8 GeV in the laboratory or .7 GeV in the center of mass.

The coincidence between two banks of scintillation counters registers muon candidates. With 4×10^{10} protons per pulse incident upon a 2% collision length target, we established that the single rate of all counters is less than 4 MC. The coincidence rate of the two scintillation counters is about 100 KC.

We have observed muons in coincidence with particles detected by the MIT-BNL double arm spectrometer. Figure 11 shows the pulse height distribution of the muon scintillation counters. Figure 12 shows the coincidence of two muon counters and Fig. 13 displays the triple coincidence between a muon and particles in each of the spectrometer arms at high beam intensities.

We estimate a hadron punch through of less than 10^{-4} and appoximately 5 in 10^4 pions decaying to contaminate the muon rate.

In the center of mass frame the detector covers a solid angle of roughly 2 s.r. The c.m.s. production angle of detected muons, θ^*, ranges from $60° - 120°$, with a minimum required momentum, p^*, of approximately 700 MeV. The acceptance of muons as functions of different values of $p_\perp^*$ and $p_\shortparallel^*$ in the cm system is shown in Fig. 14. The event rate with the muon trigger requirement is a factor of a thousand lower than that of the double arm spectrometer alone.

The ratio of the $\pi^+\pi^-$, π^-p mass spectra with/without μ is shown in Fig. 15. We see that it is flat and structureless. The same ratio for $k^-\pi^+$ channel is shown in Fig. 16. Again it is flat. In conclusion we see no structure in the reactions

$$p + Be \rightarrow h^+ + h^- + x \qquad (h^\pm = \pi^\pm, k^\pm, p^\pm)$$

in the pair mass range from 1.5 to 2.6 GeV.

If the observed direct μ's are all originated from charmed particle decay, one would expect a cross section due to charm about 10^{-29} cm^2. The upper limit we can set for the channel $k^-\pi^+\mu$ is about 10^{-31} cm^2/GeV in the mass range $m_{k\pi} = 1.5$ to 2.5 GeV, which shows that $(k^-\pi^+\mu)+x$ is not a significant decay mode of the pair of charmed particles.

ACKNOWLEDGMENT

I am grateful for valuable discussions with various members of the MIT-BNL group, especially with Professors U. Becker and S.C.C. Ting, Drs. J. Burger, J.J. Aubert, W. Toki and many others concerning the hadron-electron data and μ hadron pair data before they are published.

REFERENCES

1. J.J. Aubert et al., Phys. Rev. Lett. 33, 1404 (1974).
2. C. Bacci et al., Phys. Rev. Lett. 33, 1408 (1974).
3. J.E. Augustin et al., Phys. Rev. Lett 33, 1406 (1974).
4. G.S. Abrams et al. Phys. Rev. Lett. 33, 1453 (1974).
5. J.J. Aubert et al., Phys. Rev. Lett. 35 416 (1975).
6. J.J. Aubert et al., Phys. Rev. Lett. 35 639 (1975).

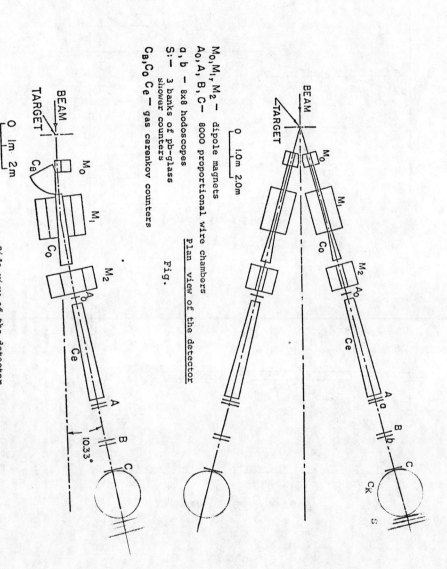

M_0, M_1, M_2 — dipole magnets
A_0, A, B, C — 8000 proportional wire chambers
a, b — 8×8 hodoscopes
S :— 3 banks of pb-glass shower counters
$C_B, C_0 \; C_e$ — gas cerenkov counters

Fig. 1 Plan and side views of the MIT-BNL double arm spectrometer set up for taking data for hadron-electron pair.

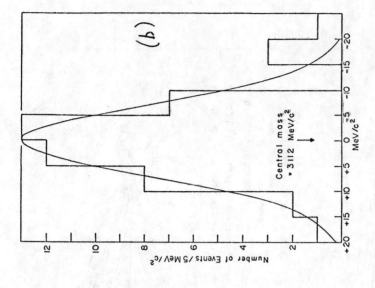

Mass spectrum for events in the mass range $2.5 < m_{ee} < 3.5$ GeV/c^2. The shaded events correspond to those taken at the normal momentum setting, while the unshaded ones correspond to a momentum setting 10% below normal. The acceptance is a smooth function of m.

Fig. 2 The J signal observed in August and October 1974 and the resolution of the spectrometer. The width of J

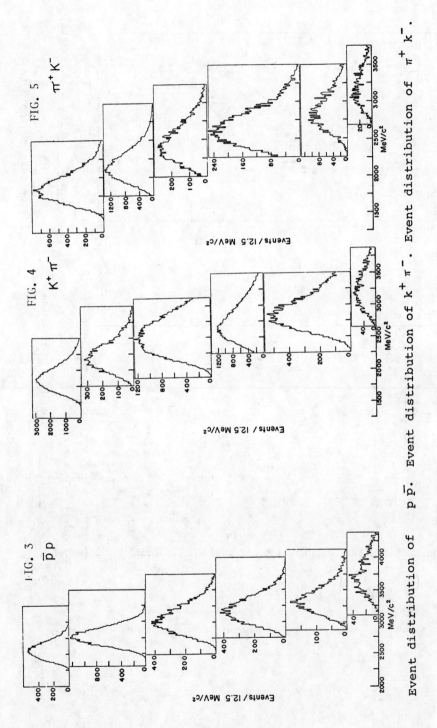

FIG. 3 $\bar{p}p$

FIG. 4 $K^+\pi^-$

FIG. 5 $\pi^+ K^-$

Event distribution of $p\bar{p}$. Event distribution of $K^+\pi^-$. Event distribution of $\pi^+ K^-$.

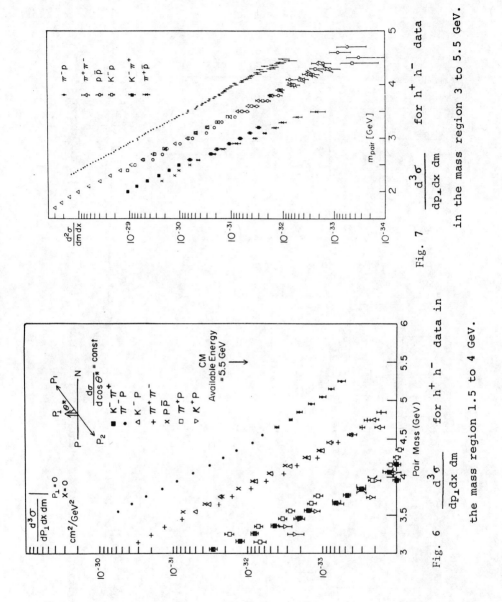

Fig. 7 $\dfrac{d^3\sigma}{dp_\perp dx\, dm}$ for $h^+ h^-$ data in the mass region 3 to 5.5 GeV.

Fig. 6 $\dfrac{d^3\sigma}{dp_\perp dx\, dm}$ for $h^+ h^-$ data in the mass region 1.5 to 4 GeV.

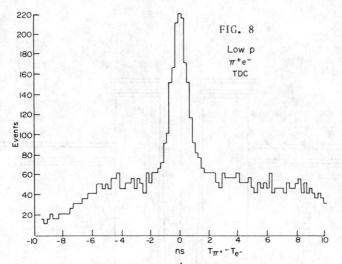

8) Time difference due to π^+ and e^- as measured in the spectrometer.

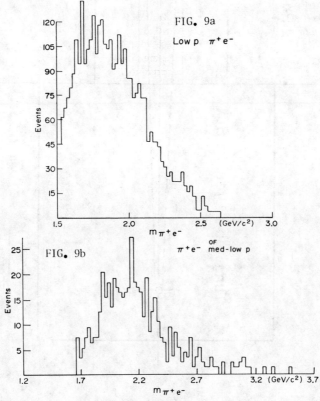

9) Event distribution of π^+e^- from 1.4 to 2.7 GeV and from 1.7 to 3 GeV.

24

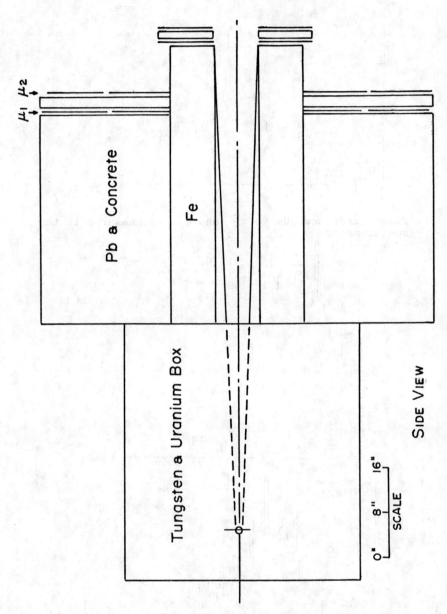

Fig. 10 Side view of the μ detector situated at the front of the spectrometer.

FIG. 11

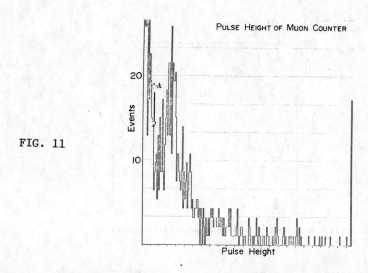

Pulse height distribution of particles going through the muon scintillation counter. One sees a peak due to relativistic charged particles and a tail due to low energy neutron or photon conversion.

FIG. 12

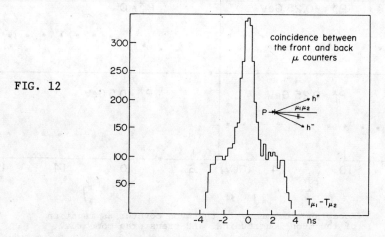

Time difference between the two muon scintillation counters. The peak corresponds to particles going through both counters at once.

FIG. 13

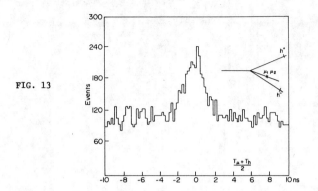

Time difference between the muon and the pair of hadrons.

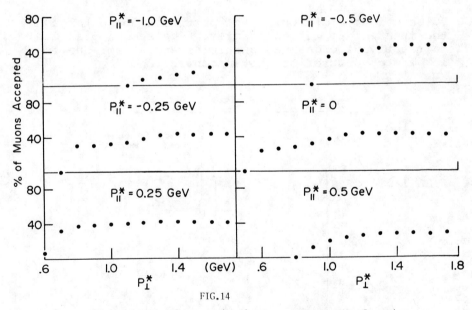

FIG. 14

Acceptance of muons in the muon counter as function of the muon longitudinal and transverse momentum.

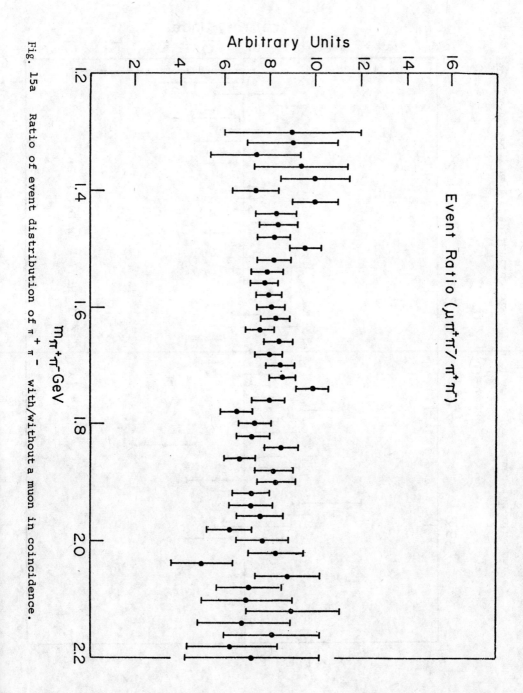

Fig. 15a Ratio of event distribution of $\pi^+\pi^-$ with/without a muon in coincidence.

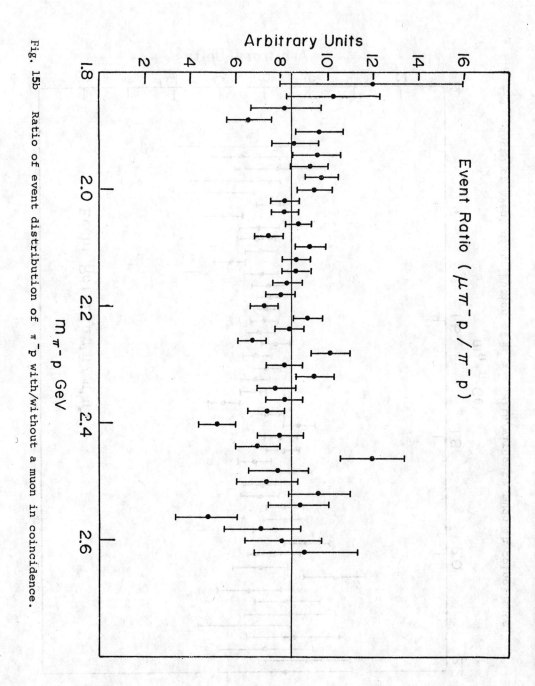

Fig. 15b Ratio of event distribution of $\pi^- p$ with/without a muon in coincidence.

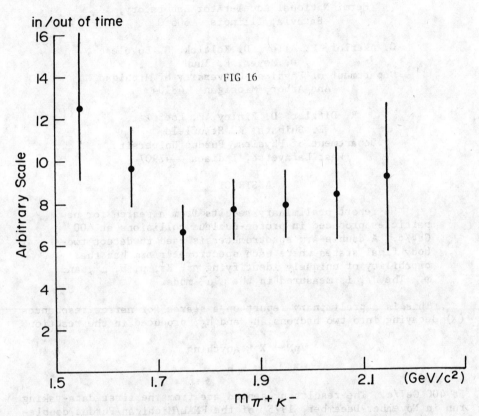

Ratio of events of $\pi^+ k^-$ with a muon in coincidence to the events without muon in coincidence as function of the pair hadron mass.

A SEARCH FOR NEW PARTICLES IN PROTON-NUCLEUS COLLISIONS AT 400 GeV/c*

D. Bintinger, D. Jovanovic, R. Lundy
Fermi National Accelerator Laboratory,
Batavia, Illinois 60510

C. Akerlof, P. Alley, D. Koltick, R. Loveless[†],
D. Meyer, R. Thun
Department of Physics, University of Michigan,
Ann Arbor, Michigan 48104

R. Ditzler, D. Finley, F. Loeffler,
E. Shibata, K. Stanfield
Department of Physics, Purdue University,
West Lafayette, Indiana 47907

ABSTRACT

We report preliminary results from a search for new particles produced in proton-nucleus collisions at 400 GeV/c. A double-arm spectrometer is used to detect two-body final states where each spectrometer arm has the capability of uniquely identifying $\pi^{\pm}$, $K^{\pm}$, p, $\bar{p}$, $\mu^{\pm}$, and φ. The J/ψ is measured in the $\mu^+\mu^-$ mode.

This is a preliminary report on a search for narrow resonances (X) decaying into two hadrons, h_1 and h_2, produced in the reaction

$$pN \rightarrow X + \text{anything}$$
$$\hookrightarrow h_1 h_2$$

at 400 GeV/c. The results presented are from the first data-taking run in November-December, 1975, of the FNAL/Michigan/Purdue double-arm spectrometer system installed in the M2 beam line at FNAL. Our experiment (E-357) is one of several performed at FNAL[1,2] to search for new particles resulting from the possible existence of a new quantum number, charm[3]. If the J/ψ particle[4] is a $c\bar{c}$ bound state, a new family of hadrons, some of which should decay into $K\pi$, K^-p, or $\bar{p}K^+$ with a significantly large branching ratio, is expected. Thus far, except for a bump in the $K^-\pi^+$ mass spectrum observed by MSU/OSU/Carleton[2], the results of charmed particle searches have been negative[1,5,6].

---

*Work supported in part by the U. S. Energy Research and Development Administration and the National Science Foundation.

[†]Present address: Fermi National Accelerator Laboratory, Batavia, Illinois 60510.

The plan view of our symmetric, double-arm spectrometer system is shown in Fig. 1. Each arm contains 16 planes of drift chambers, 20 scintillation counters, 3 Cherenkov counters, a BM-109 dipole magnet which bends vertically, and 6 feet of steel plus 18" of concrete to identify $\mu^{\pm}$. Unambiguous identification of π, K, and p and $\bar{p}$ is obtained for momenta between 7 and 20 GeV/c, and separation of p and $\bar{p}$ from π and K is obtained up to 40 GeV/c. In addition, we detect $\varphi \rightarrow K^+ K^-$, $K_s^0 \rightarrow \pi^+ \pi^-$, and $\Lambda^0 \rightarrow p\pi^-$ when their decay products go into a single arm of the spectrometer system. Thus, we can identify $X \rightarrow \pi\pi$, πK, πp, KK, Kp, pp, $\varphi\pi$, φK, and φp in all possible charge states except those with a π^0. The mass acceptance of the system for unambiguously defined particles is $1.5 \leq M_x \leq 4.0$ GeV/c^2 with the rapidity acceptance of $X \rightarrow h_1 h_2$ events confined to $y_{cm} = 0$. The mass resolution varies linearly with M_x; at $M_x = 3$ GeV/c^2, $\sigma_m \simeq 7$ MeV/c^2.

The 400 GeV/c diffracted proton beam was incident on a 10% interaction length CH_2 target divided into seven segments separated by 4" along the beam axis. The trigger for $X \rightarrow h_1 h_2$ was simply a coincidence between the two spectrometer arm signals, each of which, by the coincidence between signals from suitable sets of intra-arm scintillation counters, signified that a charged particle with momentum greater than 7 GeV/c had traversed the arm. For our typical incident flux of 5×10^7 protons per ~ 0.8 sec spill, the two arm coincidence rate was 250/spill. However, due to computer-induced deadtime, the number of events actually recorded on magnetic tape was ~ 80/spill. During the November-December, 1975, run, approximately 7.7 million triggers were recorded, with $\sim 33\%$ having reconstructable tracks in both arms.

The observation of $\varphi \rightarrow K^+ K^-$, where both the K^+ and K^- go into the same arm, provides an excellent test of our apparatus and reconstruction procedures. As shown in Fig. 2, the $\varphi \rightarrow K^+ K^-$ signal appears as a sharp peak in the $K^+ K^-$ mass spectrum with a width consistent with our calculated mass resolution. We estimate the ratio of φ plus anything in the other arm to π^- plus anything in the other arm to be about 10^{-2} for $p_\perp \gtrsim 1.4$ GeV/c. Since the branching ratio for $\varphi \rightarrow \mu^+ \mu^-$ is 2.5×10^{-4}, it is clear that φ production contributes little to the observed prompt μ to π ratio of $\sim 10^{-4}$ at large $p_\perp$.

Muon pair data were accumulated simultaneously with the hadron pair data in order to experimentally verify the mass scale, mass resolution, and sensitivity of the experiment with the J/ψ particle. Our J/ψ statistics are limited since the beam rate was optimized for hadron running. The hadron trigger rate saturated the data recording capability at $\sim 0.5 \times 10^8$ ppp. No special attempt was made to improve J/ψ statistics at the expense of the hadron data by running with only a μ-pair trigger. The μ-pair effective mass spectrum representing the data taken in November and December is presented in Fig. 3. The J/ψ signal is a sharp peak centered at a mass of 3.095 GeV/c^2. From these data we determine experimentally that our mass resolution is $\sigma_m \sim 7$ MeV/c^2. Assuming a linear dependence on atomic number, isotropic J/ψ decays, and a momentum dependence given by

$$E \frac{d^3\sigma}{dp^3} \propto (1 - |x|)^{4.3} e^{-1.6 p_\perp},$$

we calculate that $\sigma_{J/\psi} B_{\mu\mu}$ = (9 ± 3)nb/nucleon at 400 GeV/c, in agreement with Snyder et al.[7] who obtained (11 ± 3)nb/nucleon at 400 GeV/c under the same assumptions.

Convinced by our $\varphi \to K^+K^-$ and $J/\psi \to \mu^+\mu^-$ signals that we understand the apparatus well, we have investigated the hadron-pair mass spectra. In Figs. 4-8 we show the mass spectra for five of the twenty-one possible combinations of charged π's, charged K's, and p and $\bar{p}$: specifically, $\pi^+\pi^-$, $K^-\pi^+$, π^-K^+, K^-p, and $\bar{p}K^+$. Clear, narrow peaks in the $K^{\mp}\pi^{\pm}$ or K^-p and $\bar{p}K^+$ mass spectra would be indications for charmed particles. None of the mass spectra, including those not presented, show any statistically significant narrow structure at the 4 standard deviation level. Since all of these data were collected simultaneously, the $K^-\pi^+$ data may be compared with the π^-K^+ data and the K^-p with the $\bar{p}K^+$ data. In doing this, we find that none of the tantalizing, but nevertheless statistically insignificant, peaks in the $K^-\pi^+$ and K^-p spectra coincide with similarly tantalizing peaks in their respective conjugate spectra. Thus, at this early stage of our experiment we see no evidence for massive narrow resonances decaying into two hadrons.

It is clear from this experiment and others[1,2,5,6] that charmed particles are not easily observable in hadronic effective mass spectra. To obtain upper limits on the cross sections times branching ratio into two hadrons, $\sigma_c B_{h_1 h_2}$, it is necessary to make assumptions about the production mechanism for charmed particles. If we assume that charmed particles are produced with the same momentum dependence as the J/ψ particle, we can calculate upper limits for $\sigma_c B_{h_1 h_2}$ directly in units of $\sigma_{J/\psi} B_{\mu\mu}$. For the data shown in Figs. 5-8, $\sigma_c B_{h_1 h_2}$ ranges from 10 to 40 times $\sigma_{J/\psi} B_{\mu\mu}$ at M_x = 2.3 GeV/c^2 and 4 to 8 times $\sigma_{J/\psi} B_{\mu\mu}$ at M_x = 3.0 GeV/c^2. Here we have used the criterion that a 4 standard deviation peak in a 20 MeV/c^2 wide mass bin would have been a positive indication of a narrow resonance. These upper limits are set primarily by the large physical hadronic background. If we take our calculated value of (9 ± 3)nb/nucleon for $\sigma_{J/\psi} B_{\mu\mu}$, then the level of sensitivity of our particle search is of the order of 100 nb/nucleon for $\sigma_c B_{h_1 h_2}$.

Our experimental run at FNAL is now less than one-half complete. In the remainder of the run we should be able to at least double the amount of data. The anomolous lepton production observed in several diverse experiments suggests a possible signature for events containing new particles. Therefore, we have proposed an additional experiment at FNAL to search for narrow resonances in two-body hadron mass spectra for events containing a prompt muon. Only minor modifications to our present apparatus are required. This new proposal has been approved, and we hope to start taking data as early as this summer.

We are grateful for the support of the Fermi National Accelerator Laboratory staff and the technical staffs at the University of Michigan and Purdue University. We appreciate the assistance of Professor O. E. Johnson during the November-December, 1975, experimental run.

REFERENCES

1. E. J. Bleser et al., Phys. Rev. Lett. $\underline{35}$, 76 (1975).
2. J. A. J. Mathews, contribution to this Conference.
3. For a review of charm, see M. K. Gaillard, B. W. Lee, and J. L. Rosner, Rev. Mod. Phys. $\underline{47}$, 277 (1975).
4. J. J. Aubert et al., Phys. Rev. Lett. $\underline{33}$, 1404; J.-E. Augustin et al., Phys. Rev. Lett. $\underline{33}$, 1406.
5. J. J. Aubert et al., Phys. Rev. Lett. $\underline{35}$, 416 (1975); A. M. Boyarski et al., Phys. Rev. Lett. $\underline{35}$, 196 (1975).
6. N. McCubbin, contribution to this Conference.
7. H. D. Snyder et al., submitted to Phys. Rev. Lett.

PRESENT E-357 LAYOUT
FNAL/MICHIGAN/PURDUE

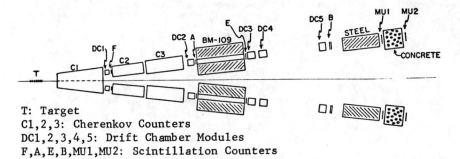

T: Target
C1,2,3: Cherenkov Counters
DC1,2,3,4,5: Drift Chamber Modules
F,A,E,B,MU1,MU2: Scintillation Counters

Fig. 1. Plan view of the double-arm spectrometer system.

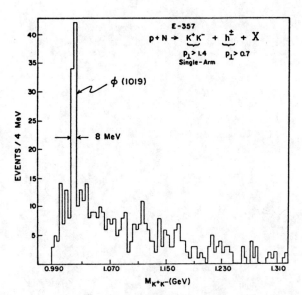

Fig. 2. K^+K^- effective mass spectrum for events in which both the K^+ and K^- go into a single spectrometer arm.

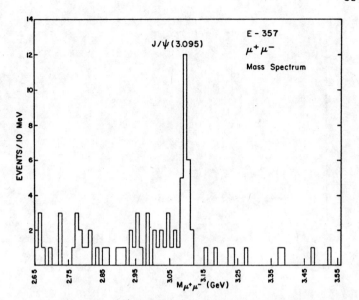

Fig. 3. $\mu^+\mu^-$ effective mass spectrum.

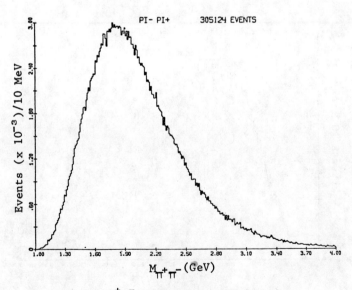

Fig. 4. $\pi^+\pi^-$ effective mass spectrum.

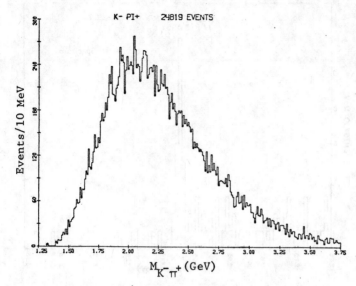

Fig. 5. $K^-\pi^+$ effective mass spectrum.

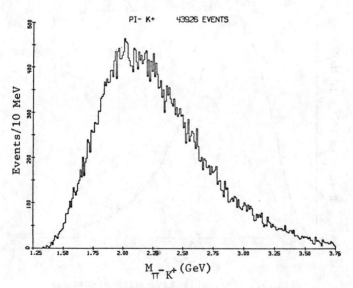

Fig. 6. π^-K^+ effective mass spectrum.

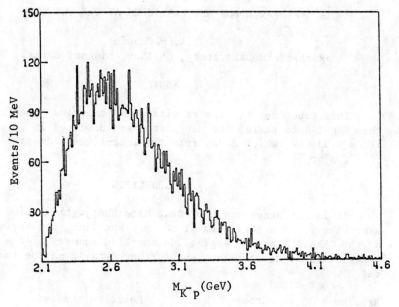

Fig. 7. K^-p effective mass spectrum.

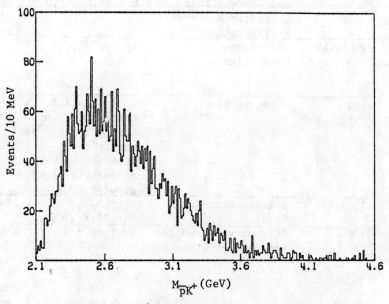

Fig. 8. $\bar{p}K^+$ effective mass spectrum.

A SEARCH FOR NEW PARTICLES AT THE CERN ISR

N.A. McCubbin
Rutherford Laboratory, Chilton, Didcot, Berks., U.K.

ABSTRACT

This paper reports the results of an experiment performed at the CERN ISR to search for new particles, produced in pp collisions at $\sqrt{s}$ = 53 GeV, which decay into two long-lived hadrons (i.e. π^+, π^-, K^+, K^-, p or $\bar{p}$).

INTRODUCTION

It is not necessary to repeat here the justification for searching for new particles at the present time, and several groups at the CERN ISR are so engaged. A one-line summary of the various experiments is given below, and further details may be found in ref. 1.

Exp. No	Group	Reaction studied	Status
R108	CERN-Columbia-Oxford-Rockefeller	pp → e^+e^- + X	Installation this summer.
R205/R206	British-Scandinavian and CERN-Holland Lancaster-Manchester collaborations	pp → $h_1 h_2$ + X h = $\pi^\pm, K^\pm$, p, $\bar{p}$	Completed. Results discussed below.
R401/R414	CERN-Hamburg-Orsay-Vienna	pp → $\mu^+\mu^-$ + X	$J/\psi(3.1)$ seen[2]. More data taking this summer.
R406	Bologna-CERN	pp → 'Exotic' + X 'Exotic' = long-lived ($\tau \gtrsim 10^{-10}$ sec) particle of fractional charge or integral charge greater than unity.	Analysis in progress.
R410/R413	British-Scandinavian-Orsay	pp → $h_1 h_2$ + X h = $\pi^\pm, K^\pm, p, \bar{p}$	Data taking
R605	CERN-Harvard-Munich-Riverside-Northwestern	pp → eμ + X	Data taking. See Rubbia's report in proceedings of this conference.

Experiment No.	Group	Reaction studied	Status
R702	CERN-Saclay-Zurich-Tel-Aviv	$pp \to e^+e^- + X$ $pp \to e\mu + X$	Data taking on e^+e^-. Installation of μ-spectrometer this summer.
R804	Frascati-Genoa-Harvard-MIT-Naples-Pisa	$pp \to \mu^+\mu^- + X$	Installation this summer.

The rest of this paper will concentrate on the results from experiment R205/R206.

RESULTS FROM THE R205/R206 EXPERIMENT

This experiment studied the reaction $pp \to h_1 h_2 + X$, where h is a $\pi^+, \pi^-, K^+, K^-, p$ or $\bar{p}$, looking for narrow enhancements in the invariant mass (M) spectrum of the two hadron system. The experiment was carried out by the British-Scandinavian and CERN-Holland-Lancaster-Manchester collaborations, and the members of these two collaborations are listed in ref. 3.

Apparatus

The apparatus consisted of two hadron spectrometers— a small angle spectrometer (SAS) positioned above the downstream arm of one of the ISR beams, and a wide angle spectrometer (WAS) which could be positioned at some chosen angle, θ, to the acute beam-beam bisector. This set-up is shown in fig. 1.

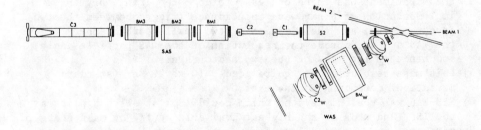

Fig. 1. Sketch of the R205/R206 apparatus showing the small-angle spectrometer (SAS) and the moveable wide-angle spectrometer (WAS).

The main characteristics of these two spectrometers are listed below:

SAS	WAS
4 magnets (S2,BM1,BM2,BM3)	1 magnet
3 Cerenkov counters (C1,C2,C3)	2 Cerenkov counters (C1W,C2W)
21 Spark Chamber planes	6 modules of spark chambers
Solid angle ~ 0.08 msr.	Solid angle ~ 11 msr.
Magnets are set to accept particles of one sign of charge (either +ive or -ive) at some central momentum p_c. Momentum bite is $\sim p_c \pm 0.5\, p_c$.	Accepts both positively and negatively charged particles with momenta $\gtrsim 0.3$ GeV/c
$\Delta p/p \lesssim 1\%$ (FWHM).	$\Delta p/p \sim 0.03\, p(\text{GeV}/c)$ (FWHM)
Particles measured: π^+, K^+, p or $\pi^-, K^-, \bar{p}$.	Particles measured $\pi^\pm, K^\pm$, p and $\bar{p}$.

Further details of these two spectrometers can be found in references 4 and 5.

The SAS was designed for a study of high x diffraction, whilst the WAS was designed for measuring high p_T production in the central region, and considered as a system for detecting the two hadron decay of some hypothetical new particle, the SAS-WAS system suffers the major draw-back of having very small acceptance. This is perhaps most dramatically seen from the fact that, at typical ISR luminosities ($L \sim 10^{31} \text{cm}^{-2} \text{sec}^{-1}$), we would expect to see one J/ψ (3.1) decay to $p\bar{p}$ every millenium or so! However on the credit side:
1) both spectrometers had been fully operational for some time (~ 3 years), were well understood, and all necessary software was written and working.
2) the experiment was therefore easy to do - requiring only the implementation of a hardware coincidence trigger between the two spectrometers, and the addition of two small spark chambers to the WAS arm to improve the resolution on the WAS particle angle and hence on the mass of the two hadron system.
3) this mass resolution was quite good - 10 to 20 MeV (standard deviation) depending on the mass.
4) SAS-WAS was the only system which could look at <u>all</u> $h_1 h_2$ pairs at the ISR, and thus at the highest available centre of mass energy.

Data taken

All data were taken at $s^{\frac{1}{2}} = 53$ GeV (26.5 GeV/c momentum protons in each ring) during the period March to July 1975, corresponding to an integrated luminosity of $8.0\; 10^{36}$ cm^{-2}. About 70% of the running time was spent with the SAS set to a central momentum of 8 GeV/c and the WAS at 38° to the acute beam-beam bisector, and the rest with the SAS central momentum at 5 GeV/c and the WAS moved out to 45°. In both cases the production angle of particles in SAS was 50 mrad. The running time was also divided into 20% of the time with positives in SAS and the rest negatives. Due to the high rate of

'leading protons' in SAS this gave roughly equal numbers of
coincidence events with positives and negatives in SAS. In terms of
the kinematic variables of the two hadron system the phase space
covered was:

$$\Delta M \sim 2 \text{ GeV (lower limit depends on } h_1 \text{ and } h_2)$$
$$x_{pair} \sim 0.2 \text{ to } 0.5$$
$$p_T^{pair} \sim 0.5 \text{ to } 1.5 \text{ GeV/c}$$

It is to be noted, however, that the acceptance of the apparatus is
such that both x and p_T are linearly related to the mass.

The gas pressures in the three SAS Cerenkov counters were
adjusted to give π,K,p identification over the entire momentum range.
The lower momentum particles in WAS - the momentum spectrum peaked
around 1 GeV/c - require a combination of time of flight and
Cerenkov analysis for particle identification, and there was a K/p
ambiguity in the region 1.4 to 1.6 GeV/c. A total of $\sim$ 900K
coincident triggers were recorded of which 170K had a reconstructed
track and unique particle identification in each spectrometer.

Results

The invariant mass spectra for h_1(SAS) h_2 (WAS) and h_1 (WAS)
h_2 (SAS) were combined where appropriate, resulting in 21 $h_1 h_2$
combinations which are listed below.

$h_1 h_2$	$\pi^+\pi^+$	$\pi^+ K^+$	$p\pi^+$	$K^+ K^+$	pK^+	pp
Q	+2	+2	+2	+2	+2	+2
B	0	0	+1	0	+1	+2
S	0	+1	0	+2	+1	0
Events	14029	2712	23998	115	1593	1761

$h_1 h_2$	$\pi^+\pi^-$	$\pi^+ K^-$	$\bar{p}\pi^+$	$K^+\pi^-$	$K^+ K^-$	$\bar{p}K^+$	$p\pi^-$	pK^-	$p\bar{p}$
Q	0	0	0	0	0	0	0	0	0
B	0	0	-1	0	0	-1	+1	+1	0
S	0	-1	0	+1	0	+1	0	-1	0
Events	52771	2668	1618	4096	304	136	26710	1261	1181

$h_1 h_2$	$\pi^-\pi^-$	$\pi^- K^-$	$\bar{p}\pi^-$	$K^- K^-$	$\bar{p}K^-$	$\bar{p}\bar{p}$
Q	-2	-2	-2	-2	-2	-2
B	0	0	-1	0	-1	-2
S	0	-1	0	-2	-1	0
Events	29989	2922	2137	81	102	40

Plotting out the 21 mass spectra in 20 MeV bins, there is no
narrow enhancement so striking as to render arguments about
statistical significance superfluous - i.e. no effect which the
community of bump-hunters would all agree was a particle and not a
statistical fluctuation. In order to analyse quantitatively the
statistical significance of the various fluctuations in the data,
we have to know, with high precision, the shape of the two hadron
mass continuum. It is of course to be expected that this continuum

will provide a substantial background to any resonance signal
because of the high charged hadron multiplicity produced by pp
collisions at ISR energies. (Indeed this is one of the reason why
several groups have chosen to look at leptons where the naturally
occurring 'background' from pp collisions is much smaller). We have
obtained this continuum spectrum for any h_1 (SAS) h_2(WAS) combination
by combining every h_1 4-vector measured in SAS with every h_2 4-vector
measured in WAS, and calculating the resultant mass spectrum. So
for example if we have 10^4 π^+ in SAS, produced in coincidence with
any hadron in WAS, and 10^4 π^+ in WAS, produced in coincidence with
any hadron in SAS, we obtain a continuum mass spectrum for the
combination $\pi^+\pi^+$ containing 10^8 pseudo events. Spectrometer
acceptance effects, and also the average effect of the small
kinematic correlation resulting from the coincidence requirement,
are automatically incorporated in the mass spectrum obtained. We
then simply normalize this spectrum to the number of real
coincidence events, and calculate the signed standard deviation, χ,
for each bin from $\chi = (n_d - n_c)/n_c^{\frac{1}{2}}$ where n_d is the number of the real
data events and n_c is the number in the normalized generated spectrum.

Eight of the twenty-one mass spectra are shown in figs. 2 to 9.
(All the mass spectra will be published in the very near future). In
each plot the normalized continuum is plotted through the data and
the signed standard deviation for each bin is plotted below, where
necessary combining bins until the population ≥ 5.

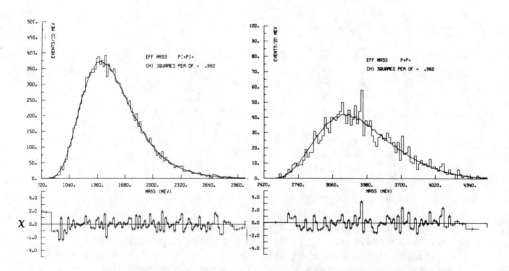

Figs. 2 and 3. Mass spectra for $\pi^+\pi^+$ (fig. 2) and pp (fig. 3), with
the normalized generated continuum (see text)
superimposed, and the signed standard deviation, χ,
plotted below.

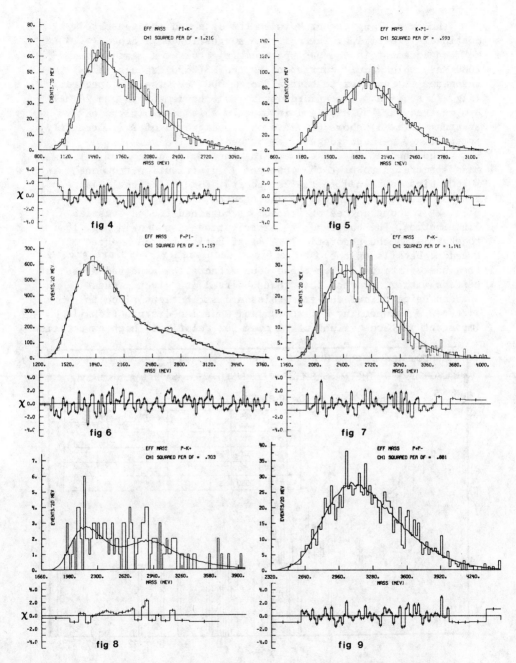

Figs. 4 to 9. Mass spectra for $\pi^+K^-, \pi^-K^+, p\pi^-, pK^-, \bar{p}K^+, p\bar{p}$ (figs. 4 to 9 respectively) with the normalized generated continuum (see text) superimposed, and the signed standard deviation, χ, plotted below.

The overall agreement between the data and the generated continuum is striking. There is the suggestion of a broad ($\sim$ 100 MeV) enhancement in the $p\pi^-$ spectrum (fig. 6) at a mass around 2680 MeV, which could correspond to the N*(2650) though this resonance is reported to have a Γ of $\sim$ 300 MeV. The pK^- spectrum (fig. 7) shows a 3.4 standard deviation enhancement in one 20 MeV bin centred at 2150 MeV, and in exactly this mass bin the $\bar{p}K^+$ spectrum (fig. 8) shows 6 events on a background of 2.7. Certainly interesting, perhaps suggestive, but not conclusive!

A convenient way of summarizing the entire data from all 21 mass spectra is to plot the observed χ^2 distribution from the $\sim$ 1600 bins contained in the 21 spectra. For sufficiently large bin populations, χ^2 should be distributed as $(2\pi\chi^2)^{-\frac{1}{2}} \exp(-\chi^2/2)$ -i.e. χ^2 for one degree of freedom as obtained from a Gaussian distribution. The observed χ^2 distribution is shown in fig. 10 together with the theoretical χ^2 distribution for one degree of freedom. Details of the four effects seen with $\chi^2 > 10$ are given in the inset of fig. 10. For these four effects the expected number of effects with χ^2 greater than that observed are given - these numbers being calculated from Poisson distributions since the difference between the Poissonian and Gaussian distributions is important at large values of χ^2 even for relatively high expectation

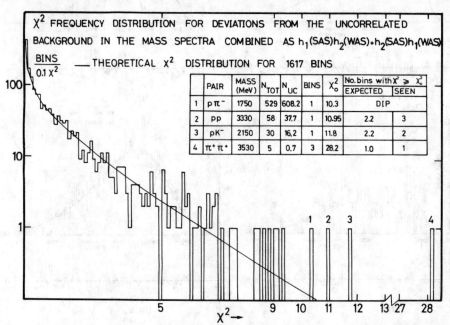

Fig. 10. Observed χ^2 distribution for the 1617 bins comprising the 21 $h_1 h_2$ mass spectra. Superimposed is the theoretical χ^2 distribution for 1 degree of freedom.

values. It is clear from fig. 10 that all the fluctuations seen are quite compatible with statistics. Indeed, as assurance that we understand what we are doing, the agreement between the observed χ^2 distribution and the theoretically expected one is rather comforting!

It is of course possible that a new narrow resonance might exist in several charge states, nearly degenerate in mass, which decay into different $h_1 h_2$ combinations. Accordingly we have combined the original 21 mass spectra into 5 mass spectra distinguished only by baryon number, $B = -2,-1,0,+1,+2$. The B = 2 and +2 are just the $\bar{p}\bar{p}$ and pp spectra of the original set of 21, and the other three are shown in figs. 11,12 and 13. The χ^2

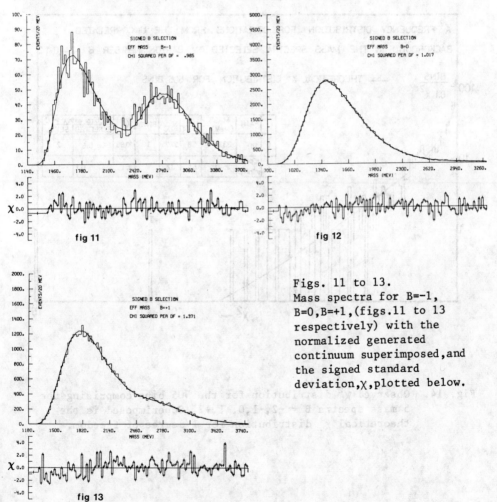

Figs. 11 to 13. Mass spectra for B=-1, B=0, B=+1, (figs.11 to 13 respectively) with the normalized generated continuum superimposed, and the signed standard deviation, χ, plotted below.

distribution for the entire data sample combined in this way is shown in fig. 14. A 3.9 s.d. enhancement is seen in the B = +1 channel at a mass of 3510 MeV. It thus arises from combining the $p\pi^+, p\pi^-, pK^+$, and pK^- channels. The region around 3510 MeV of the B = +1 spectrum is shown in more detail in fig. 15, and the same mass region for the four contributing $h_1 h_2$ combinations in figs. 16,17,18 and 19. As a statistical fluctuation such an enhancement is a 1 in 10 chance (see inset of fig. 14), and so the evidence for a new particle at this mass, whilst suggestive, is not conclusive.

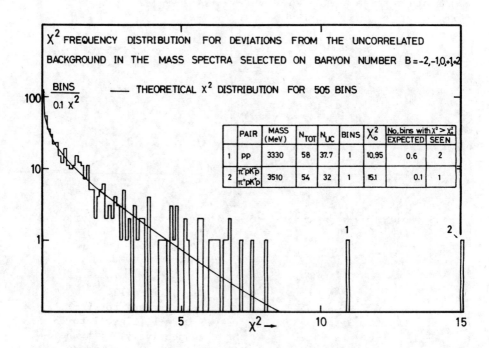

Fig. 14. Observed χ^2 distribution for the 505 bins comprising the 5 mass spectra B = -2,-1,0,+1,+2. Superimposed is the theoretical χ^2 distribution for 1 degree of freedom.

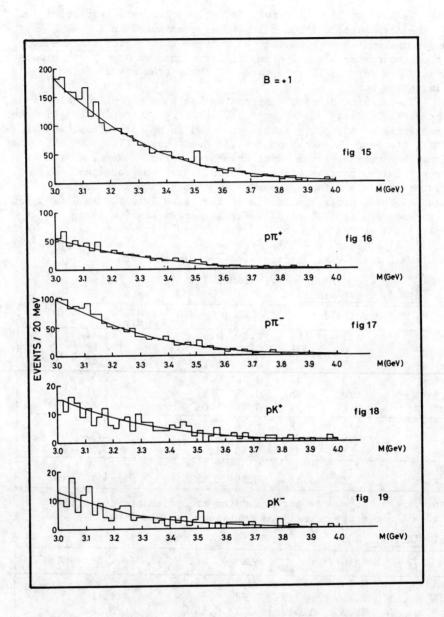

Fig. 15 to 19. Mass spectra for B = +1 (fig. 15) and $p\pi^+, p\pi^-, pK^+, pK^-$ (figs. 16 to 19 respectively) in the mass region around 3510 MeV. The normalized generated continuum is superimposed on the data.

We have tried making cuts on various kinematic variables and on the associated charged multiplicity, measured in a $\sim 4\pi$ hodoscope system surrounding the intersection region, in the hope of enhancing the signals seen or uncovering new ones, but to no avail. The very small acceptance of the system was a serious limitation here.

The acceptance of the apparatus has been calculated allowing for particle decay and absorption. Assuming isotropic decay in the rest frame of any new particle produced in pp collisions, we can integrate over the decay angles and quote upper limits on the invariant cross-section times branching ratio. Several of these upper limits, corresponding to 5 s.d. above the continuum, are listed in Table I for various h_1 h_2 combinations and various masses. These upper limits lie in the range 1 to 10 μb GeV^{-2}. For reference the inclusive π^+ invariant cross-section in pp collisions at $\sqrt{s} = 53$ GeV is ~ 10 mb GeV^{-2} at $x \sim 0.3$ and $p_T = 0.4$ GeV/c. Also given in Table I are the values of the invariant cross-section for the various 'signals' seen in the data.

TABLE I

a) 5 s.d. upper limits

h_1 h_2	Mass(GeV)	$n_{5\ s.d.}$*	x	p_T GeV/c	$E\frac{d\sigma}{d^3p} \cdot$ BR μb GeV^{-2}
π^+ π^-	1.5	187	0.31	0.65	13.75
π^+ π^-	2.0	96	0.35	0.9	4.18
π^+ π^-	2.5	39	0.4	1.15	1.40
K^+ π^-	1.5	37	0.32	0.6	22.43
K^+ π^-	2.0	45	0.33	0.8	6.18
K^+ π^-	2.5	24	0.4	1.05	1.45
K^- π^+	1.5	38	0.27	0.6	6.06
K^- π^+	2.0	29	0.3	0.8	2.12
K^- π^+	2.5	15	0.43	1.12	0.92
pK^-	2.5	27	0.34	0.71	7.04
pK^-	3.0	18	0.39	0.9	1.16

*n_{5sd} = No. of events corresponding to 5 standard deviations above the normalized continuum level.

b) 'Signals'

h_1 h_2	Mass(GeV)	n_{signal}*	x	p_T GeV/c	$E\frac{d\sigma}{d^3p} \cdot$ BR μb GeV^{-2}
pK^-	2.15	13.8	0.3	0.55	19.55
$\bar{p}K^+$	2.15	3.3	0.29	0.56	1.16
$p\pi^-$	2.68	110.5†	0.39	0.85	8.58
B = +1					
$p\pi^+, p\pi^-$ pK^+, pK^-	3.51	22	0.45	1.0	0.23

*n_{signal} = data - normalized continuum
†Total signal in six 20 MeV bins.

Kinoshita et al.[6] have calculated the invariant cross-section for charmed meson production (D^o and $\bar{D}^o$) in pp collisions at $\sqrt{s}$ = 53 GeV. Their model gives a good description of the inclusive π and K spectra. Our upper limits for D^o and $\bar{D}^o$ production, assuming a 50% branching ratio into $K\pi$[7] and extrapolating to p_T = 0.4 GeV/c using exp $(-2.2\ p_T)$ as proposed by Kinoshita et al., are shown in fig. 20. These upper limits lie about an order of magnitude above the theoretical curves.

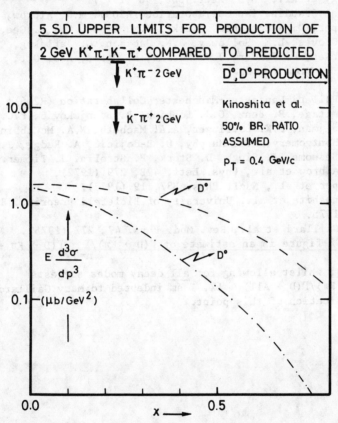

Fig. 20. 5 s.d. upper limits for D^o and $\bar{D}^o$ production compared to theoretical predictions of Kinoshita et al.

Conclusions

The SAS-WAS experiment has yielded no conclusive evidence for a new narrow resonance decaying into two long-lived hadrons. However the upper limits obtained leave room for charm production at a level predicted by (at least some) theorists.

The results reported here represent the combined efforts of

the R205/R206 collaborations. It is a pleasure to acknowledge in
particular the unstinting work of Gerjan Bobbink in the final data
analysis stage.

REFERENCES

1. Experiments at CERN in 1975 (CERN-RD/52-1000-August 1975).
2. E. Nagy et al., Phys. Lett. $\underline{60B}$, 96 (1975).
3. British-Scandinavian Collaboration (R205): M.G. Albrow,
   B. Alper, D. Aston, A.G. Clarke, W.M. Evans, C.N.P. Gee,
   E.S. Groves, L.E. Holloway, J.N. Jackson, G. Jarlskog,
   H. Jensen, D. Korder, J.V. Morris, H. Ogren, R.J. Ott,
   S. Rock, W.A. Wenzel.

   CERN-Holland-Lancaster-Manchester Collaboration (R206):
   J. Armitage, P. Benz, G.J. Bobbink, B. Bosnjakovic, F.C. Erné,
   P. Kooijman, F.K. Loebinger, A.A. Macbeth, N.A. McCubbin,
   H.E. Montgomery, P.G. Murphy, D. Radojicic, A. Rudge, J.C. Sens,
   A.L. Sessoms, J. Singh, D. Stork, P. Strolin, J. Timmer.
4. M.G. Albrow et al., Phys. Lett. $\underline{42B}$, 279 (1972).
5. B. Alper et al., Nucl. Phys. $\underline{B87}$, 19 (1975).
6. K. Kinoshita et al., University of Bielefeld preprint, Bi-75/12,
   June 1975.
7. M.K. Gaillard et al., Rev. Mod. Phys. $\underline{47}$, 277 (1975).
   The 50% figure is an estimate of $\Gamma(D \to \overline{K\pi})/\sum_{n=0} \Gamma(D \to K\pi + n$ soft

   pions), whilst allowing for <u>all</u> decay modes suggests
   $\Gamma(D \to K\pi)/\Gamma(D \to All) \sim 5\%$. I am indebted to Mary Gaillard for a
   clarification of this point.

CHARMED PARTICLE SEARCHES USING BUBBLE CHAMBERS WITH HADRONIC BEAMS

R. Harris
Fermi National Accelerator Laboratory, Batavia, Illinois 60510

ABSTRACT

Searches for charmed particle production using bubble chambers with hadronic beams are summarized. These searches depend on the detection of neutral strange particles. Upper limits are given for long-lived particles ($t_c > 10^{-11}$ sec), for short-lived particles showing up in mass distributions and for semi-leptonic decays involving μ's plus V's.

Bubble chambers have several important advantages over counter experiments in searches for charmed particles. They are 4π detectors - all charged tracks are seen and the biases are relatively small. Phase space restrictions are minor; in the large bubble chambers the full allowed range in x and $p_\perp$ is seen. The bubble chamber is also a good detector of neutral strange particles, through their characteristic "V" decays. Since charmed particles are expected to decay into strange particles, the detection of one or more neutral strange particles can be used as a signal to search for charm.

Recent searches for charm using hadronic beams are listed in Table I. The various methods used, and the resultant 95% confidence level upper limits are summarized in Table II. There are two basic types of search, depending on whether the charmed particle is assumed to be long-lived ($t_c \gtrsim 10^{11}$ sec), or short-lived. To search for long-lived particles, which would be visible in the bubble chamber, one looks for even-prong decays of neutral particles or odd-prong charged decays. An early search of this type,[6] in pp interactions at 400 GeV/c, yielded upper limits of $\gtrsim 20$ μb. The FSU[2] and Fermilab-FSU[3] groups have searched for such particles in two-prong V decays not fitting K_s^0, Λ, or $\bar{\Lambda}$. These are candidates for $M_c^0 \to K\pi$. Out of ~10,000 V's, the FSU group has found one candidate, whose mass as $\pi^+\pi^-$ is 1.832 ± .035 GeV. With the tracks assumed to be $K^{\pm}\pi^{\mp}$, the mass is 1.926 ± .037 GeV or 1.988 ± .036 GeV. If one assumes, however, that two such events must be found to indicate a signal, this sets an upper limit using Poisson statistics of 0.7 μb. The Fermilab-FSU group[3] has made a similar search at higher energy of 5200 V's and found no clean candidates with a Kπ mass above 1.5 GeV, indicating an upper limit of 4 μb.

Searches for short-lived charmed particles consist of searching for narrow resonances in mass distributions, such as $K_s^0(n\pi)$ $\Lambda(n\pi)$, $\bar{\Lambda}(n\pi)$ and Kp, with up to 6 pions. The Columbia group[1] reports upper limits in the range 1.5-3.7 μb for several exclusive channels, such as $\pi^+ p \to B_0^{++} + \bar{M}_0^0 + \pi^+$. These were determined by combining all possible decay modes of the two charmed particles and examining scatter plots of the resulting mass distributions. Inclusive channels

TABLE I: Experimental Parameters

Group	Interaction		E_{CM}	Number of Events
1) Columbia-Binghampton[1]	$\pi^+ p$	15 GeV/c	5.4 GeV	15,233 K^o, Λ
2) FSU-Brandeis[2]	$\pi^+ d$	15 GeV/c	7.7 GeV	~10,000 V's
3) Fermilab-FSU[3]	$\pi^- p$	250 GeV/c	21.7 GeV	~700 $K^o, \Lambda, \bar{\Lambda}$
4) Davis-Krackow-Seattle-Warsaw[4]	$\pi^- d$	200 GeV/c	27.5 GeV	72 μ triggers
5) Santa Cruz-SLAC[5]	$\pi^+ N$	15 GeV/c	5.4 GeV	12,000 μ triggers

have been searched by examining mass distributions for all possible $V^o(n\pi)$ combinations. Representative upper limits are given in Table II; these upper limits are strongly dependent on mass of the combination, as illustrated in Fig. 1 for $K_s^o \pi^+$ from the Fermilab-FSU data.

Attempts to enhance the charm signal have been made in these searches by applying various cuts to the data. No charmed signal has been found by this procedure but the upper limits are reduced, as illustrated in Table III from the Fermilab-FSU data. Cuts in $p_\perp$, also made by the FSU group,[2] are an attempt to remove background from unassociated back-to-back particles coming from the two different fragmentation cones. The large number of background combinations inherent in the mass distributions can be reduced if one assumes charm decays only to low multiplicity states. The Fermilab-FSU group[3] attempts to enhance their signal by using the fact that the Fermilab 15' bubble chamber can detect with reasonable efficiency events with two neutral strange particles. If the charm-anticharm pair both decay into neutral strange particles these events will contain an enriched charm signal. The background in this case can be further reduced by requiring that the $V^o \bar{V}^o$ mass be greater than 1.5 times threshold, thus eliminating threshold $V^o \bar{V}^o$ production. These various cuts reduce the upper limits considerably, as shown in Table III.

Two further experiments have been done to detect semi-leptonic decays of charmed particles. In each case a muon detector was used to tag the decay. In the Davis-Krackow-Seattle-Warsaw experiment,[4] the Fermilab 30" bubble chamber was used to search for V decays in conjunction with the μ's. The number of V^o's per event was found to be similar with or without the μ trigger. Unfolding the detection efficiency using a Monte Carlo, this leads to upper limits for $V^o + \mu$ in the range 300-500 μb. The Santa Cruz-SLAC experiment[5] measured

TABLE II: Types of Search and Upper Limits (95% CL)

A. Long-Lived States ($t_c \gtrsim 10^{-11}$ sec)

   1) $V^o \to K^{\pm}\pi^{\mp}$ search:

   - Michigan-Rochester[6]: $\sigma \lesssim 20$ μb

   - FSU[2]: $\sigma \lesssim 0.7$ μb

   - Fermilab-FSU[3]: $\sigma < 4$ μb

B. Short-Lived States

   1) Exclusive channels, (i.e., $\pi^+ p \to p + M_c^+ + M^o$)

   - Columbia[1]: $\sigma \lesssim 3.7$ μb (See Ref. 1)

   2) Inclusive channels: $C \to K\pi$, $K\pi\pi$, $\Lambda\pi$, $\Lambda\pi\pi$, etc.

   Representative Upper Limits:

Channel	Columbia[1]	FSU[2]	Fermilab-FSU[3]	UCSC-SLAC[5]
	All	$p_\perp$ cut	Double Strange	μ trigger
$K_s^o \pi^+$ (2.0 GeV)	10 μb	4 μb	28 μb	5 μb
$\Lambda \pi^+ \pi^+$ (2.5 GeV)	6.5 μb	2 μb	42 μb	4 μb
$\bar{\Lambda}\pi^+$ (2.5 GeV)	-	-	24 μb	-

C. Semi-Leptonic Decay: μ + V

   1) Davis-Krackow-Seattle-Warsaw[4]: $\sigma_{lim} \sim 300$–500 μb

all μ triggered events with the SLAC streamer chamber. These events have been searched for possible hadronic decays of a charmed particle in conjunction with the assumed semi-leptonic decay of the other charmed particle. Using inclusive mass distributions, upper limits of 1-11 μb have been determined for a number of decay states.

Clearly, charm has not been found in these experiments. The upper limits determined here can be compared with various expectations of the charm particle cross section. Two types of theoretical models have been applied to charm production. For diffractive production of charm at large x, triple Regge models[7] predict $\sigma(c\bar{c}) \sim 1$ μb at Fermilab energies. These predictions depend strongly on energy. In Fig. 2, several curves from the central production model of Einhorn and Ellis[8] are shown along with the upper limits described above. The curves, which represent different assumed gluon distributions, give the total $c\bar{c}$ production cross section, while the experimental values give ranges of cross section times branching ratio. The theoretical predictions indicate that the expected $\sigma(c\bar{c})$ is much larger at Fermilab energies than at the lower energies. However, considerably more statistics would be needed before any of these experiments reached the theoretical limits. Two published counter experiments are shown for comparison;[9] the Northwestern-Rochester experiment is sensitive only to $x > .3$ and thus should be compared to theoretical limits of ~1 nb. Finally, an experimental indication of the expected cross section can be found from the fact that the deuteron-antideuteron cross section is ~1 μb at Fermilab energies. Thus the existence of charm is not yet ruled out by the experiments described here.

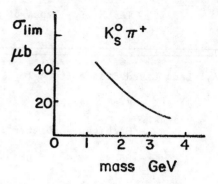

Fig. 1: 95% CL Upper Limit Fermilab-FSU Data

TABLE III: Effect of Cuts on Upper Limits

Fermilab-FSU[3] 95% CL Upper Limits (μb)

	Mass	All	$p_\perp > 1$ GeV	$n_c \leq 6$	Double Strange	Threshold*
$K_s^o \pi^+ \pi^-$	2.0 GeV	165	50	54	91	65
$\Lambda \pi^+$	2.5 GeV	42	28	22	25	21
$\Lambda \pi^+ \pi^+$	2.5 GeV	80	31	19	42	30

*Double Strange, with $m(VV) > 1.5$ x threshold

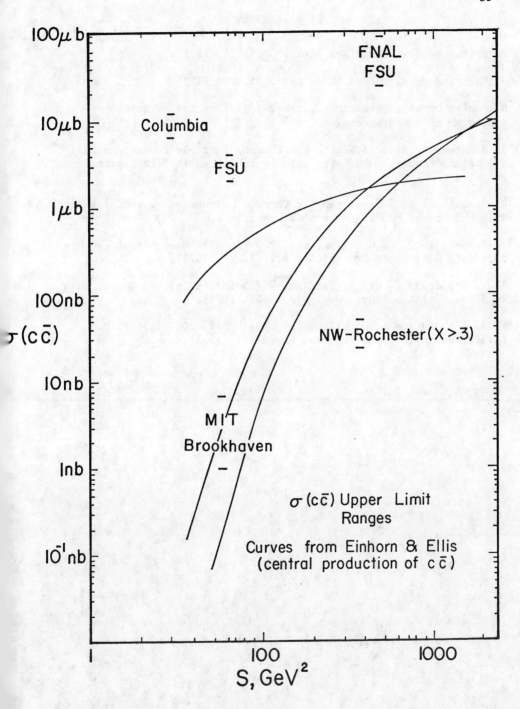

Fig. 2: σ(c c̄) Upper Limits

REFERENCES

[1] C. Baltay et al., Phys. Rev. Lett. $\underline{34}$, 1118 (1975).

[2] V. Hagopian et al., Phys. Rev. Lett. $\underline{36}$, 296 (1976).

[3] D. Bogert et al., Fermilab Note FN-281 (1975), and material presented to this Conference.

[4] J. H. Klems et al., "A Search for Charmed Particle Production in π^-d Interactions at 200 GeV/c", to be published in Acta Physica Polonica.

[5] K. Bunnell et al., "A Search for Charmed Hadrons Using a Direct-Muon Trigger", paper presented to this conference.

[6] T. Ferbel, in D. A. Garelick, ed., Experimental Meson Spectroscopy-1974 (AIP Conference Proceedings #21, N. Y., 1974), p. 393.

[7] R. D. Field and C. Quigg, Fermilab-75/15-THY (1975); V. Barger and R. J. W. Phillips, Phys. Rev. $\underline{D12}$, 2623 (1975).

[8] M. B. Einhorn and S. D. Ellis, Phys. Rev. $\underline{D12}$, 2007 (1975), D. Sivers, Rutherford Lab Preprint RL-75-171 (1975).

IMPLICATIONS FOR NEW PARTICLE SEARCHES OF SOME COSMIC RAY EMULSION EVENTS

T. K. Gaisser
Brookhaven National Laboratory, Upton, NY 11973[†]
and
Bartol Research Foundation of The Franklin Institute[*]
Swarthmore, PA 19081

ABSTRACT

We summarize calculations of backgrounds for cosmic ray candidates for new particles in emulsions. Even if only the strongest cosmic ray candidates are accepted as genuine, the implication is that their production cross sections would be relatively large at accelerator energies ($\gtrsim 1\mu b$) and that decay modes with π^0 appear to be prominent. We refer to current accelerator upper bounds on $\sigma \times B_{K\pi}$ for production of hadrons with $M > 2$ GeV and discuss the relation between these bounds and the cosmic ray events.

Several candidates for new, short-lived stable particles with masses of order 2 GeV have been found in interactions of ~10 TeV cosmic rays.[1] A schematic diagram of one such event[2] is shown in Fig. 1. Another candidate for pair production of new particles has been reported recently by Sugimoto et al.[3] Both putative new particles (X) are charged and their decay modes appear to be $X^{\pm} \to h^{\pm} + \pi^0$ and $X^{\pm} \to h^{\pm} + \eta$, where $h^{\pm}$ is an unidentified charged particle that is not an electron. The masses and lifetimes of X depend on the identification of $h^{\pm}$. Results for various possibilities are summarized in Table I. As the search at accelerators for charmed or other new particles with lifetimes in the range 10^{-14}-10^{-12} sec intensifies, a re-evaluation of these cosmic ray events is in order. This is particularly so since the high spatial resolution of emulsions makes them appropriate for detecting tracks with lifetimes in this range; indeed emulsion experiments are underway at FNAL.

With this motivation, F. Halzen and I have made a detailed calculation[4] of backgrounds that could simulate production and decay of a new particle in emulsions and emulsion chambers. The principal sources of background we have considered are
 a) Diffraction dissociation of a secondary prong on a target nucleus in the detector,
 b) Decay of an ordinary hyperon, and
 c) Elastic or quasi-elastic scattering of a secondary prong.
Processes (a) and (b) can simulate both $X^{\pm} \to h^{\pm}$ + missing neutrals

[†] Work supported, in part, by Energy Research and Development Administration.
[*] Work supported, in part, by National Science Foundation. Permanent Address.

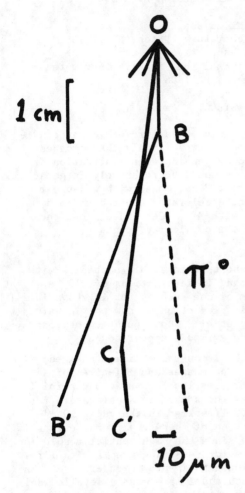

Fig. 1 A schematic drawing of the original Niu event[2] showing possible pair production at O of particles OB and OC. Track OB apparently decays at B into a charged particle BB' and a π^0. Track OC may decay at C into charged particle CC' and missing neutrals. The original interaction at O produced 70 charged tracks and 23 high energy γ's.

and $X \to h_1 + h_2$, where both decay products are seen. (Most of the candidates summarized in Ref. 1 show only two-body decay.) Process (c) is a background only for $X^{\pm} \to h^{\pm} +$ missing neutrals. In practice, ordinary hyperon decay is negligible in cases where two decay products are seen and the invariant mass of the parent X can be estimated.

Results of the background estimates are presented in detail in Ref. 4 and summarized here in Table II. The basic qualitative result is that these backgrounds are smaller than most accelerator physicists had thought, and that events in which both candidates for an associated pair are seen have a low probability of being background. It is relevant to mention in this connection that we have assumed for our calculations the estimates quoted by the authors[2,3] for accuracy of vertex location, energy, angle, and coplanarity of X, h_1 and h_2. We have further assumed that all tracks are associated with the same event.

If all sixteen candidates from the six cosmic ray interactions summarized in Ref. 1 are indeed new hadrons then
i) $\sigma_X \times B_{\text{2-body}} \sim 1/3$mb at 10 TeV,
ii) The new particles are frequently produced in the beam fragmentation region,
iii) Production is usually in events with unusually high multiplicity, and
iv) The branching ratio for two-body decays is large, and these decays often involve a π^0.
Items (ii)-(iv) may be partly the result of the bias introduced by scanning for high energy γ rays. If so, this fact would only increase estimates of σ_X. These consequences (together with the related fact that large σ_X requires

Table I Masses in GeV (and lifetime in sec) of possible new particles for various assumptions about the unidentified charged decay product, $h^{\pm}$.

$h^{\pm}$	Niu, et al. (Ref. 2) $X^{\pm} \to h^{\pm} + \pi^0$	Sugimoto et al. (Ref. 3)	
		Prong 2 $X^{\pm} \to h^{\pm} + \eta$	Prong 20 $X^{\pm} \to h^{\pm} + \pi^0$
π	1.79 ± 0.17 (2.2×10^{-14})	1.50 ± 0.38 (4.6×10^{-13})	1.59 ± 0.40 (3.1×10^{-12})
K	2.15 ± 0.20 (2.6×10^{-14})	1.66 ± 0.42 (5.1×10^{-13})	1.74 ± 0.44 (3.4×10^{-12})
P	2.95 ± 0.26 (3.6×10^{-14})	1.98 ± 0.50 (6.1×10^{-13})	2.10 ± 0.53 (4.1×10^{-12})
Σ	3.50 ± 0.30 (4.3×10^{-14})	2.23 ± 0.56 (6.8×10^{-13})	2.36 ± 0.59 (4.5×10^{-12})

correspondingly small (< 1%) branching ratio to leptons to avoid conflict with prompt lepton data) led us to conclude in Ref. 4 that it is unlikely that all of the cosmic ray candidates are new particles.

If, on the other hand, only one (or both) of the cleanest events (e.g., Ref. 2 or 3) involves production of a new hadron then one would guess that the estimate (i) would not be reduced by more than an order of magnitude. The accuracy of such a conjecture is limited by standard considerations about statistics of small numbers and by lack of information about the totality of events scanned in the cosmic ray searches. (Our estimate in (i) of 1/3mb is based on the statement of Ref. 1 that the production rate of new particles is about one per hundred interactions.)

Nevertheless, it seems plausible to take 30μb as a nominal value for a cosmic ray estimate for $\sigma_X \times B_{2\text{-body}}$ at 10 TeV for the purposes of comparison with accelerator searches for charm. One can make such a comparison with the help of the excitation curve for $p\bar{p}$ or $K\bar{K}$ production and an assumption of s/M^2 scaling.[5] For production of a pair of particles each with M = 2 GeV the production cross section should increase by a factor of ten between 200 and 10^4 GeV. This leads to an estimate of 3μb for $\sigma_X \times B_{2\text{-body}}$ at 200 GeV. In contrast, upper limits on $\sigma_{charm} \times B_{K\pi}$ presented at this conference[6,7] are about one order of magnitude less than this, of order

Table II   Summary of Backgrounds

Signature	B/interaction	B	Source
(one-prong or with π^0)	$\sim 10^{-4}$	~ 0.01	Diffraction Dissociation
(kink)	$\gtrsim 10^{-3}$	$\sim 10\%$	Diffraction Dissociation + hyperon decay
	$\gtrsim 2 \times 10^{-3}$	$\sim 20\%$	elastic scatter
(e.g. Ref. 2)	few $\times 10^{-7}$	few $\times 10^{-5}$	All sources w/coplanarity
	few $\times 10^{-6}$	few $\times 10^{-4}$*	without coplanarity
(e.g. Ref. 3)	$\sim 10^{-8}$	$\sim 10^{-6}$	All sources w/coplanarity
	$\sim 10^{-6}$	$\sim 10^{-4}$*	without coplanarity

* The larger number is obtained when the strict coplanarity condition of Ref. 2 is relaxed. It gives a more conservative estimate of the background for the Niu event.[2] For the conditions of Ref. 3 only the larger number is relevant. (See Ref. 4 for further discussion of coplanarity.)

several hundred nanobarns, for $M_{K\pi} > 2$ GeV.

The accelerator limits and the cosmic ray observations are not in conflict, however, because

1) The accelerator limits for $\sigma \times B_{h_1 h_2}$ concentrate on the mass range $M_{h_1 h_2} > 2$ GeV, whereas the cosmic ray estimates of M_X can be significantly less than 2 GeV, especially when the unidentified charged decay product is a pion (see Table I).

2) Moreover, accelerator limits on $\sigma_{charm} \times B_{\pi\pi}$ tend to be higher than for $\sigma_{charm} \times B_{K\pi}$.

It seems fair to conclude, however, that, if the events of either Ref. 2 or Ref. 3 (or both) involve production of new particles with large two-body decay modes these ought to be seen very shortly in accelerator searches.

Finally, one should not overlook the possibility of some less orthodox interpretation of events such as those of Refs. 2 and 3. For example, if the coplanarity condition is relaxed, the observed decays could include a missing ν. Such events could result from production and decay of massive, integrally charged fermions, such as the unconfined quarks of Pati and Salam[8] or the F of Feinberg and Lee.[9] Because of the missing ν, these would be hard to see in present accelerator searches that look for bumps in two-hadron invariant mass distributions.

ACKNOWLEDGMENT

I am grateful to Francis Halzen for reading the manuscript, and to E. Shibata and J. Matthews for helpful conversations.

REFERENCES

1. K. Hoshino et al., Proc. 14th International Cosmic Ray Conference (Munich) 7, 2442 (1975) and references quoted therein.
2. K. Niu et al., Progr. Theoret. Phys. (Kyoto) 46, 1644 (1971). It is interesting to note that this event was interpreted immediately as production of charm; T. Hayashi et al., Progr. Theoret. Phys. (Kyoto) 47, 280 (1972).
3. H. Sugimoto et al., Progr. Theoret. Phys. (Kyoto) 53, 1541 (1975).
4. T.K. Gaisser and F. Halzen, Nucl. Phys. (submitted) Jan. 1976.
5. T.K. Gaisser and F. Halzen, Phys. Rev. D11, 3157 (1975).
6. E. Shibata, these Proceedings.
7. J. Matthews, these Proceedings.
8. See, e.g., J.C. Pati, A. Salam and S. Sakakibara, preprint (Jan. 15, 1976).
9. G. Feinberg and T.D. Lee, preprint CO-2271-74 (1976).

COMMENTS ON THE NEW PARTICLES*

Thomas Appelquist[+]
Yale University, New Haven, Connecticut 06520

ABSTRACT

The spectroscopy and decays of the new particles are reviewed from the point of view of the charmonium model. The review proceeds from general features expected to be true in a wide class of spin 1/2 constituent models to predictions that depend in more detail on the dynamics of the color SU(3) quark gluon theory. A list of important and unresolved problems is presented at the end.

I. INTRODUCTION

The growing body of experimental data[1,2,3] on the ψ particles (J,ψ',...) increasingly points to their interpretation as bound states of a heavy spin 1/2 particle bound to its antiparticle. The system bears a striking resemblance to positronium, but with a much larger scale for interaction[4] strengths and binding energies. A specific theory of the ψ particles with these qualitative features[5] identifies the constituent fermions as one or more heavy quarks with forces mediated by the exchange of massless vector gluons in a color-SU(3) gauge theory[6] of strong interactions. This specific theory is what I mean by the charmonium model, having adopted the generic name charm[7,8] to denote the new quantum number(s) of the heavy quark(s). In reviewing the model, I will generally be proceeding from features likely to be true in a large class of models with spin 1/2 constituents to predictions that depend on more detailed uses of the color SU(3) dynamics or the weak interaction properties of the new quark(s).

Sections II through VII cover the main topics:

II. Level Structure and Transitions—Qualitatively

III. The J and the η_c

IV. ψ' Decays and Intermediate Levels

V. Structure above the ψ'

VI. The OZI Rule

I will try at the end (Section VII) to summarize what

I consider to be the outstanding problems--both experimental and theoretical.

There are several features of the color SU(3) quark-gluon theory to keep in mind as background for the topics I will discuss.

1. Quarks come in three colors and a variety of flavors: u,d,s,Q,.... The lightest of the new quarks may or may not be the charmed quark of GIM[8] with charge 2/3 and preferential weak coupling to strangeness.

2. Forces arise from the exchange of colored gluons which couple <u>universally</u> to each quark flavor. The universality is enforced by the local gauge invariance.

3. The theory is renormalizable and because of asymptotic freedom[9], the short distance behavior is computable to all orders in g. Above an energy scale of about 1 Gev, the effective coupling strength is small ($g_{eff}^2/4\pi \ll 1$) and perturbation theory can be applied.

4. At energy scales smaller than about 1 Gev, the effective coupling strength grows and perturbation theory becomes useless. This infrared growth may result in the permanent confinement of quarks and gluons to the interior of color singlet hadrons. Although it has not been demonstrated, I will assume permanent quark confinement as one of the features of the charmonium model.

The role of large and small energy scales is very important in the charmonium model. Since the momentum transfers within the ψ particle turn out to be on the order of a few hundred Mev, it is quite clear that perturbation theory can not be applied. A phenomenological potential motivated by work on quark confinement replaces perturbation theory in most computational work[10,11]. This is a sensible thing to do since the constituent motion in the ψ particles appears to be non relativistic.

II. LEVEL STRUCTURE AND TRANSITIONS--QUALITATIVELY

If a spin 1/2 particle is bound to its antiparticle by a strongly confining potential (as opposed to say a Coulomb potential) and the motion is non relativistic, the spectrum of low lying states is shown qualitatively in Fig. 1.

The identification with the ψ states begins with the J(3095) as the $1\ ^3S_1$ and the ψ'(3684) as the $2\ ^3S_1$. It would seem as though three intermediate levels have so far been reached by photon emission from the ψ'.[2,3] There is a natural spin parity state $\chi(3.41)$ along with apparently two states at around 3.53 Gev. I will call them $\chi(3.51)$ and $\chi(3.55)$. It is natural to identify the $\chi(3.41)$ and $\chi(3.51)$ with the $^3P_0$ and $^3P_1$ states. Simple models of the spin forces then place the $^3P_2$ state somewhat above 3.61 and it presumably has not yet been seen. The identification of the $\chi(3.55)$ with the $2\ ^1S_0$ (η_c') state is suggested by its preference to decay directly into hadrons rather than into J(3095)+γ. This E1 transition is suppressed by the wave function overlap integral just as ψ' (3684)$\rightarrow\eta_c+\gamma$ is suppressed. The expected $1\ ^1S_0$ (η_c) state may also have been discovered at about 2.8 Gev.[2,3]

The qualitative picture looks quite good for the states below 3.7 Gev. The situation with respect to the higher states (both predicted and observed) is much less clear and I will postpone discussion to section V. They are located very near or above some sort of threshold, which in the charmonium model corresponds to the production of charmed hadrons. At the moment, the problems with the lower states are quantitative—the sizes of splittings, total widths, and transition widths. Before discussing these problems, it is important to emphasize the extreme fragility of even the qualitative picture. Unless a one to one correspondence between predicted and discovered levels continues to emerge, the simple picture of a single spin 1/2 particle bound to its antiparticle will have to be abandoned.

III. THE J AND THE η_c

There is good evidence that the J acts as an I=0, G=1 object in its direct decay into hadrons. The SU(3) character of the J is less clear. However, since the decay modes forbidden to an SU(3) singlet are not seen,[12] I will assume it to be a singlet. In the charmonium model, the constituents themselves are SU(3) singlets.

Most phenomenological work has included a long range spin-independent piece in the potential of the form ar (linearly rising) to represent the quark confining forces. With an assumed electric charge of 2/3 for the heavy quark, a fit to the leptonic width

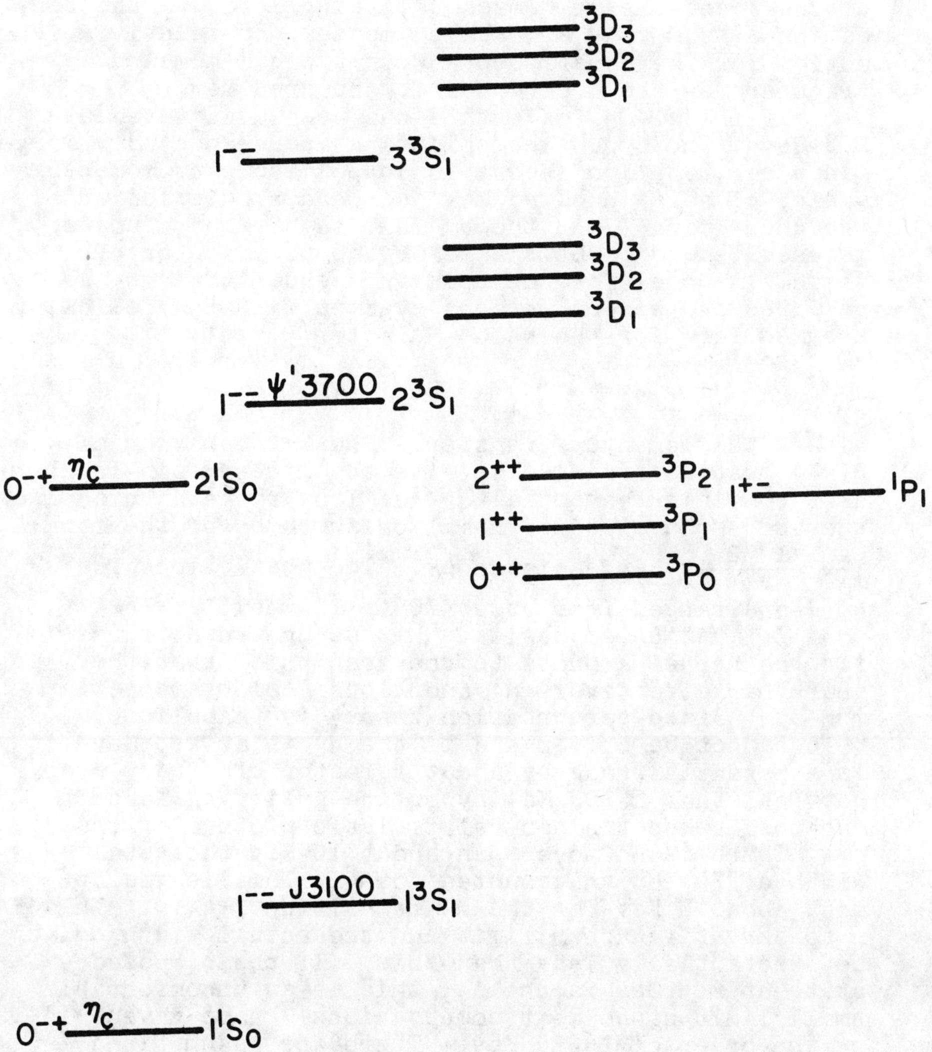

                                              ——— ³D₃
                                              ——— ³D₂
                                              ——— ³D₁

                          1⁻⁻ —————— 3³S₁

                                              ——— ³D₃
                                              ——— ³D₂
                                              ——— ³D₁

                 1⁻⁻ —ψ'3700— 2³S₁

    0⁻⁺ —η'_c— 2¹S₀   2⁺⁺ —————— ³P₂   1⁺⁻ —————— ¹P₁
                           1⁺⁺ —————— ³P₁
                           0⁺⁺ —————— ³P₀

                 1⁻⁻ —J3100— 1³S₁

    0⁻⁺ —η_c— 1¹S₀

Fig. 1   Level structure in the charmonium model.

of the J and the J-ψ' mass splitting leads to the condition $a \ll m_Q^2$.[5,10,11] This implies non-relativistic motion for the constituents with typical momentum transfers on the order of a few hundred Mev.

If the $1\,^1S_0$ (η_c) state has been discovered at 2.8 Gev[2], the hyperfine splitting is unexpectedly large. Since perturbation theory is inapplicable at momentum scales of a few hundred Mev, no hard prediction from the underlying field theory is possible but educated guesses tended to be smaller by about a factor of three. A model for the spin independent forces is required and several groups started with the following assumed form for the Bethe-Salpeter kernel:

$$K = [\gamma_\mu]_1 \, D(k^2) \, [\gamma^\mu]_2 \,, \qquad (1)$$

with k the momentum transfer. The $k \to 0$ behavior of $D(k^2)$ must be arranged to give the linear spin-independent confining potential ar at long range. This fixes the spin-spin piece of Hamiltonian to be of the form $\frac{a}{m_Q^2} \frac{\sigma_1 \cdot \sigma_2}{r}$ at large distances. The predictions for $\Delta m(J-\eta_c)$ ranged from about 70 to 100 Mev.[13]

It must be emphasized that using tensor form (1) for the kernel amounts to the assumption that the Q quark has effectively no anomalous "color" magnetic moment. Since perturbation theory is inapplicable, this cannot be proven and if the η_c is at 2.8 Gev, it is apparently wrong by about a factor of $\sqrt{3}$.[14] I do not feel that a 300 Mev hyperfine splitting forces one to give up the non-relativistic picture of the J.

There is no more than about 10% of the total J width of 70 Kev unaccounted for[12]. This leaves at most about 7 Kev for the M1 transition $J \to \eta_c + \gamma$. This is a conservative upper limit and the actual width could be substantially less than this. If the proposed state at 2.8 Gev is the η_c, this is an uncomfortably small limit since most computations[15] have given widths on the order of 20-30 Kev. There are assumptions involved in the computations (such as a normal Dirac magnetic moment and charge 2/3 for the Q quark) but without reducing the quark charge, it seems very difficult to bring the result down to a few Kev or less. This may develop into one of the most serious problems for the model.

One other piece of experimental information on this dipole transition has been reported from DESY[2,3]. If the reported state at 2.8 Gev is the η_c, then

$$\frac{\Gamma(\eta_c \to \gamma\gamma)}{\Gamma(\eta_c \to \text{all})} \times \frac{\Gamma(J \to \eta_c \gamma)}{\Gamma(J \to \text{all})} \sim 1.5 \times 10^{-4} \,. \qquad (2)$$

Two branching ratios are involved and the first is an important handle on the dynamical basis for the OZI rule. I will return to a discussion of the significance of Eq. 2 in section VI.

IV. ψ' DECAYS AND INTERMEDIATE LEVELS

The total width of the ψ' is about 220 Kev but apart from gamma ray decays into c=+ intermediate states, only about 70% of this width has been accounted for. This includes the cascade to the J (∼57%), the direct decay into hadrons ($\lesssim$10%), the decay into hadrons through a virtual gamma (∼3%) and the decay into e^+e^- or $\mu^+\mu^-$ (∼2%).[12]

The direct decay width of the ψ' ($\lesssim$20 Kev) is less than half the hadronic width (∼50 Kev) of the J. This could mean that the direct decay of the loosely bound $Q\bar{Q}$ system into hadrons is a relatively local process, measured essentially by the square of the wave function at the origin $|\psi(0)|^2$. Recall that since $\Gamma(\psi' \to e^+e^-) \approx 1/2.3\ \Gamma(J \to e^+e^-)$, it is clear that $|\psi_{\psi'}(0)|^2 \lesssim 1/2 |\psi_J(0)|^2$. I will return to a discussion of the local annihilation picture in section VI.

If some of the intermediate χ levels are P states, then their approximate location is in good agreement with the linear potential model. The χ(3.41) level has natural spin parity, having been seen in the decay and ψ'→γ+hadrons at SLAC[2] and in the ψ'→J+2γ chain decay at DESY[2]. It is natural to assign it to the $J^{PC}=0^{++}$, $^3P_0$ state of charmonium. The χ(3.51) level has also been seen in both these decays but the absence of the ππ and $K\bar{K}$ modes in its hadronic decay points to its assignment as the $J^{PC}=1^{++}$, $^3P_1$ charmonium state. If these assignments are accepted, then the existing models of spin dependent forces predict that the $J^{PC} = 2^{++}$, $^3P_2$ state should be located above 3.6 Gev[15]. For example, Schnitzer[15] predicts

$$\frac{\Delta M(2^{++},1^{++})}{\Delta M(1^{++},0^{++})} = 1.2 - 1.4 \ . \quad (3)$$

While the absolute magnitude of the spin dependent forces may be impossible to predict (as indicated by the η_c), the above ratio is probably on more solid ground. If so, the 2^{++} state has not yet been seen.

The broad peak at 3.53 seen at SLAC in the decay ψ'→γ+hadrons may, as we have already conjectured, be composed of two levels, χ(3.51) and χ(3.55). The upper level is not seen in the cascade decay ψ'→J+2γ.

This suggests that it could be the η_c' state $2\,{}^1S_0$ with $J^{PC} = 0^{-+}$. The orthogonality of the radial wavefunctions of the η_c' and J would suppress the second stage of the decay just as the E1 transition $\psi' \to \eta_c + \gamma$ is suppressed. If this assignment is correct, then $\Delta M(\psi', \eta_c') \approx 130$ Mev. This number is no easier to predict than $\Delta M(J, \eta_c)$ but if the $\Delta M(J, \eta_c)$ is 300 Mev as indicated, then a reduction by about a factor of two for the radially excited system seems sensible.[16]

The E1 transitions from the ψ' to the P states have been computed by the Cornell group.[11] They incorporate the effect of the coupling of these states to the nearby decay channels which serves to suppress the E1 transitions. If the ${}^3P_0$ state is identified with $\chi(3.41)$, they find $\Gamma(\psi' \to {}^3P_0 + \gamma) = 36$ Kev. The SLAC-LBL group is now quoting a branching ratio of 8-10% for this decay,[17] i.e. a width of about 20 Kev. The E1 decay into the $\chi(3.51)$, ${}^3P_1$ state is predicted to occur with a width of about 10 Kev and the M1 decays $\psi' \to \eta_c + \gamma$ and $\psi' \to \eta_c' + \gamma$ with widths of 10 and 2 Kev respectively. As these radiative widths are measured, they could well begin to fill out the unaccounted for fraction ($\approx 25\%$) of the total ψ' width.

None of the M1 or E1 transitions seems now to be in glaring disagreement (by, say, an order of magnitude) with the computations. Although most of the measured widths do seem to be a bit low, I don't think it is yet time to abandon the charge 2/3 assignment for the constituents.

V. STRUCTURE ABOVE THE ψ'

In the energy region above the ψ', several interesting things happen.

1. The total hadronic cross section increases. $R(E_{CM})$, the ratio of the total cross section to the $e^+e^- \to \mu^+\mu^-$ cross section changes from $R(E_{CM}) \approx 2.5$ for $E_{CM} \lesssim 3.5$ Gev to $R(E_{CM}) \gtrsim 5$ for $E_{CM} \gtrsim 5$ Gev.

2. The total cross section develops a rather broad resonance structure in the 4.1 and 4.4 Gev regions. The properties of the 4.4 resonance have been reported at this conference.[18] The mass is 4.414 Gev, the total width is about 30 Mev and the lepton pair width is about 400 ev. The 4.1 Gev region appears to contain at least two and perhaps three peaks.

3. At or above center of mass energies of 4 Gev, events of the form $e^+e^- \to e^\mp \mu^\pm$ + (≥ 2 unobserved particles) have been observed.[19]

The parton model computation of the total cross section[20] predicts a constant value for $R(E_{CM})$ away from threshold regions. In the four quark model with charge 2/3 for the fourth quark, $R(E_{CM}) \approx 3\,1/3$ above about 5 Gev. The simplest way out of this problem is the addition of more quarks and/or heavy leptons. Heavy leptons could provide an economical explanation of both the total cross section and the μ-e pair events. If more quarks are operative, then one must explain the absence of further narrow ($\lesssim 1$ Mev) structure in the total cross section. If only one heavy quark is operative, then probably two heavy charged leptons are required to understand the total cross section behavior.[20]

The dynamics of the transition region (3.7 Gev-5 Gev) appears to be more complicated than anticipated. With only one heavy quark, no more than three peaks in this region were predicted. One would be the third $^3S_1$ state and the other two would be $^3D_1$ resonances. All would be broadened by being above charm threshold. It now appears that the structure in this region is much richer than this.[18] Two proposals have already been made to explain this structure in terms of exotic quark configurations.[21]

A complete departure from the confined quark plus heavy lepton scheme has recently been proposed by several groups.[22,23] These authors suggest that the ψ constituents are not confined and are, through their weak decay, the parent particles in the μ-e pair events.[19] The most dramatic prediction of this kind of model concerns the energy dependence of the μ-e pair events. Although no explanation is offered for the total cross section structure in the 4.1 and 4.4 regions, this structure is presumably associated with the production of the ψ constituents. Since they are the source of the μ-e pair events, the cross section for these events should show the same structure as the total cross section. The measurement of this energy dependence is _very_ important. From a purely phenomenological point of view, the unconfined constituent model may appear more natural. It is certainly less contrived than placing a heavy lepton threshold nearly on top of charm threshold. On the other hand, if it is correct, we may be rather far from an understanding of the strong interactions.

VI. THE OZI RULE

The narrowness of the ψ particles is explained, within the quark model, by the Okubo-Zweig-Iikuka (OZI) rule.[24] The dynamical origin of this rule has not been established but the fact that it is so strongly operative in the new, heavy particles suggests that asymptotic freedom may play some role. This suggestion[4] has been discussed and reviewed several times[25,26] and the only thing I want to do in this talk is to remind you of the essential idea and its most immediate experimental test.

The $Q\bar{Q}$ pair in the ψ particles can annihilate through gluons into ordinary hadrons. In the case of $J(\eta_c)$ the minimum number is three (two). If the perturbation expansion for the decay matrix elements converges, then the minimal gluon mechanism will dominate and the widths can be computed. The convergence has been discussed and checked through lowest order[25] but it has not been proven. The main prediction is that the η_c should be approximately one hundred times as broad as the J--on the order of a few Mev. Other tests of this idea have been listed by Harari[26] but the η_c width is the most direct.

With this width the branching ratio for $\eta_c \to \gamma\gamma$ should be about 10^{-3}.[25] If the reported state at 2.8 Gev[2,3] is the η_c, then an interesting constraint, given by Eq. 2, already exists on this branching ratio. If the $\eta_c \to \gamma\gamma$ branching ratio is indeed on the order of 10^{-3}, this constraint tells us that the $J \to \eta_c + \gamma$ branching ratio is on the order of 10^{-1}. This is a possible and not unreasonable value. If, on the other hand, the $\eta_c \to \gamma\gamma$ branching ratio is much larger, then the $J \to \eta_c + \gamma$ branching ratio must become quite small, perhaps too small for a 300 Mev splitting.

VII. SUMMARY

There are a variety of experimental questions which must be answered and which probably can be answered in the near future.

1. The level structure and quantum numbers of the intermediate states must be pinned down. Have three levels $\chi(3.41)$, $\chi(3.51)$ and $\chi(3.55)$ already been found? Can they be identified with the $^3P_0$, $^3P_1$ and $2\ ^1S_0$ (η_c') charmonium states? If the η_c is at 2.8 Gev, then this is a reasonable mass for the η_c'. Is the $^3P_2$ located above 3.6 Gev as indicated by simple models of the spin-dependent forces? If it is not too close to or above the ψ',

it can perhaps be seen in the E1 transition $\psi' \to {}^3P_2+\gamma$. Along with the C=+1 intermediate levels, one C=-1 level, ${}^1P_1$ is expected at around 3.5 Gev.[5,15] One way to get at this state might be through the decay of one of the states in the 4.1 region into ${}^1P_1+2\pi$.

2. Is the $1\,{}^1S_0(\eta_c)$ the reported level at 2.8 Gev? A smaller hyperfine splitting (<100 Mev?) would perhaps be easier to understand but then surely the $\eta_c'-\psi'$ splitting would be at least this small. If that is the case, the η_c' has not yet been seen and the identity of the $\chi(3.53)$ states remains a puzzle.

3. Are the E1 and M1 transition widths compatible with charge 2/3 for the constituents? The E1 transition $\psi' \to {}^3P_0+\gamma$ may now be coming into line.[11,17] Will this transition and the transition $\psi' \to {}^3P_1+\gamma$ make up a good part of the ψ' width? The M1 transition $J \to \eta_c+\gamma$ may become a real problem. A width of a few Kev or less seems very unnatural with a 300 Mev splitting. Integral charge for the constituents, as suggested by several authors[22] would exacerbate these problems even more.

4. Do the J and ψ' constituents couple universally to other quarks? The charmed particles must be found. If, in the region where R is rising and falling, the μ-e pair events show no structure, then the J constituents are probably not the progenitors of these events.[22] They, of course, cannot be if they are confined heavy quarks. The structure in R at charmed particle threshold remains to be fully explored and explained.

5. If nature is good enough to provide us with very heavy quarks (m>8-10 Gev), then true charmonium could be discovered at PEP. All momentum scales, including the momentum transfer in the ground state, are large (>1 Gev) and perturbation theory can be applied to ground state properties. A hyperfine splitting of about 50 Mev is expected for 8 Gev quarks. If the J-η_c splitting is indeed 300 Mev, this is an especially striking prediction. It is also a prediction on rather solid ground. The

use of perturbation theory should be beyond reproach for this problem.

6. Does local annihilation through a minimum number of gluons have anything to do with the decay of the ψ particles? The crucial test is the width of the η_c. Is it on the order of a few Mev? If so, the $\gamma\gamma$ branching ratio should be about 10^{-3} and the product of branching ratios (Eq. 2) reported from DESY, assuming the η_c is at 2.8 Gev, can probably be lived with. If the η_c is narrower, then the $\eta_c \to \gamma\gamma$ branching ratio is larger and the M1 transition $J \to \eta_c + \gamma$ must be extremely small. On the theoretical side, the minimal gluon mechanism must be explored more carefully. It is not yet completely clear whether the use of perturbation theory can be justified for this important problem.[25]

*Research (Yale Report COO-3075-141) supported in part by the U.S. Energy Research and Development Administration.

+Alfred P. Sloan Foundation Fellow.

REFERENCES

1. J.J. Aubert et al., Phys. Rev. Lett. <u>33</u>, 1404 (1974); J.E. Augustin et al., Phys. Rev. Lett. <u>33</u>, 1406 (1974); G.S. Abrams et al., Phys. Rev. Lett. <u>33</u>, 1453 (1974).
2. W. Braunschweig et al., Phys. Lett. <u>57B</u>, 407 (1975); Phys. Rev. Lett. <u>35</u>, 1323 (1975); B. Wiik, <u>Proceedings of the 1975 International Symposium on Lepton and Photon Interactions at High Energies</u>, W.T. Kirk, editor (Stanford Linear Accelerator Center, Stanford, California 1976), pg. 69; G. Feldman, ibid., pg. 39; G.J. Feldman et al., Phys. Rev. Lett. <u>35</u>, 821 (1975).
3. Invited talks at this conference by C. Friedberg (LBL), K. Pretzl (DORIS, DASP) and M. Breidenbach (SLAC).
4. T. Appelquist and H.D. Politzer, Phys. Rev. Lett. <u>34</u>, 43 (1975).
5. Many of the qualitative features of the ψ system were predicted and outlined in two papers:

E. Eichten et al., Phys. Rev. Lett. $\underline{34}$, 369 (1975) and T. Appelquist et al., Phys. Rev. Lett. $\underline{34}$, 365 (1975).

6. H. Fritzsch, M. Gell-Mann and H. Leutwyler, Phys. Lett. $\underline{47B}$, 365 (1973).

7. B.J. Bjorken and S.L. Glashow, Phys. Lett. $\underline{11}$, 255 (1964).

8. S.L. Glashow, J. Iliopoulos and L. Maiani, Phys. Rev. $\underline{D2}$, 1285 (1970).

9. H.D. Politzer, Phys. Rev. Lett. $\underline{30}$, 1345 (1973); D.J. Gross, F. Wilczek, Phys. Rev. Lett. $\underline{30}$, 1343 (1973).

10. B. Harrington, S.Y. Park and A. Yildiz, Phys. Rev. Lett. $\underline{34}$, 168 (1975). H. Schnitzer and J.S. Kang, Phys. Rev. D12, 841 (1975). K. Jhung, K. Chung, and R.S. Willey, University of Pittsburgh preprint (1975).

11. E. Eichten et al., Phys. Rev. Lett. $\underline{36}$, 500 (1976) and references therein.

12. For a recent review, see F. Gilman, Invited Talk at the Orbis Scientiae 1976, University of Miami, Coral Gables, Florida, January 19-22, 1976.

13. H.J. Schnitzer, Phys. Rev. D13, 74 (1976).

14. For a recent critical discussion of $Q\bar{Q}$ fine structure effects in the color gauge theory, see A. Duncan, Institute for Advanced Study preprint, December, 1975.

15. J. Borenstein and R. Shankar, Phys. Rev. Lett. $\underline{34}$, 619 (1975); J. Pumplin, W. Repko and A. Sato, Michigan State preprint, July 1975; H.J. Schnitzer, Phys. Rev. Lett. 35, 1540 (1975).

16. J. Daboul and H. Kraseman, DESY preprint DESY 75/46, November 1975.

17. C. Friedberg, Invited talk at this conference.

18. M. Breidenbach, Invited talk at this conference.

19. M.L. Perl et al., Phys. Rev. Lett. $\underline{35}$, 1489 (1975).

20. E.C. Poggio, H.R. Quinn and S. Weinberg, Harvard University preprint, December 1975.

21. M. Bander et al., UC-Irvine preprint No. 75-54, 1975 (unpublished); C. Rosenzweig, University of Pittsburgh preprint PITT-158, 1975 (unpublished).

22. Boris Arbuzov, Gino Segre and Jacques Weyers, CERN preprint TH-2122-CERN, January 1976; S. Nussinov, R. Raitio and Matts Roos, SLAC preprint, SLAC-PUB-1690, December 1975; G. Feinberg and T.D. Lee, Columbia preprint, CO-2271-74, January 1976.

23. J. Pati, A. Salam and S. Sakakibara, University of Maryland preprint, January 1976.

24. S. Okubo, Phys. Lett. $\underline{5}$, 165 (1963); G. Zweig (1964), unpublished; J. Iizuka, Supplement to

the Progress of Theor. Phys. <u>37-38</u>, 21 (1966).
25. T. Appelquist and H.D. Politzer, Phys. Rev. D<u>12</u>, 1404 (1975); T. Appelquist, "Gauge Fields and Strong Interactions", Fermilab-Lecture-75/02-THY September, 1975.
26. H. Harari, <u>Proceedings of the 1975 International Symposium on Lepton and Photon Interactions at High Energies</u>, W.T. Kirk, editor (Stanford Linear Accelerator Center, Stanford, California 1976).

ψ AND EXCESS LEPTONS IN PHOTOPRODUCTION*

D. M. Ritson[†]
Stanford University, Stanford, CA 94305

ABSTRACT

The A-dependence of ψ photoproduction was measured on Beryllium and Tantalum. From this we found $\sigma_{\psi N}$ = 2.75 ± 0.90 mb. A study was made of excess leptons relative to pion production in photoproduction. A μ/π ratio of 1.40 ± 0.25 x 10^{-4} was found at 20 GeV incident photon energy. The energy dependence of ψ photoproduction was determined and appeared to have a "pseudo-threshold" at 12 GeV.

INTRODUCTION

This is a preliminary report on an experiment run at the SLAC linear accelerator in Fall 1975. The experiment was motivated by the observation[1] that ψ mesons can be detected via their two lepton decay mode by both coincidence detection of the lepton pair and by detecting one lepton in the kinematically enhanced region corresponding to $p_\perp$ = $M_\psi/2$.

We report the following measurements:

a. The energy dependence of the ψ photoproduction cross section. Results from the 1975 SLAC Photon Conference indicated a strong change in cross section in the neighborhood of 12 GeV. However, data below 11.5 GeV came from a Cornell experiment and above 13 GeV from a SLAC experiment.

b. A direct measurement of the ψ-nucleon total cross section made by determining the A-dependence of the diffractive photoproduction cross section on Beryllium and Tantalum. Previous indirect arguments using vector dominance indicated a total cross section of ~ 1 mb.

c. For photoproduction experiments in contradistinction to hadron production the "excess-lepton" signal is dominated by ψ decays at $p_\perp$ ~ 1.5 GeV/c. This made it interesting to investigate prompt lepton production in photoproduction to see whether other sources of prompt leptons such as "charm" might also show up more clearly in photon reactions.

---

*Work supported by the U. S. Energy Research and Development Administration.

[†]The work described here was done in collaboration with U. Camerini, J. G. Learned, R. Prepost, and D. E. Wiser, University of Wisconsin; R. L. Anderson, W. W. Ash, D. B. Gustavson, T. Reichelt, D. J. Sherden and C. K. Sinclair, Stanford Linear Accelerator Center.

APPARATUS

The experiment was run using the SLAC End Station A Spectrometer Facility with the standard photon and electron beams. The 8 and 20 GeV/c spectrometers were used to detect muons and pions by means of Cerenkov counters and iron range telescopes. Hodoscope or wire chamber information was not needed for these measurements. The layout of the target area is shown in Fig. 1.

A variety of remotely selectable solid targets was available, including Tantalum, Aluminum, 0.5", 2.0" and 4.0" thick Beryllium, and null targets. The targets were enclosed in a narrow helium-filled scattering chamber with thin aluminum windows. The chamber was made long enough to keep the beam entry and exit windows out of the field of view of the spectrometers. The entire target assembly could be remotely moved from side to side in order to reduce the target material traversed by a particle originating along the beam path as it traveled toward one spectrometer or the other. Since the spectrometers were on opposite sides of the beam line, the target position could only be optimized for one spectrometer at a time. Care was taken to ensure that the beam was fully within the target and not spilling over the edge. The target selection and coordinate were recorded by the computer.

A helium-filled duct downstream from the target reduced room background from the beam. Photon beams were stopped by a secondary emission quantameter (SEQ) within a concrete shielding enclosure at the rear of the end station, which measured the total energy in the beam. Electron beams passed through the end station to a remote shielded beam dump, and were measured by the SLAC integrating toroid monitors upstream of the target. One toroid, ahead of the photon beam production target, was also useful for monitoring the photon beams.

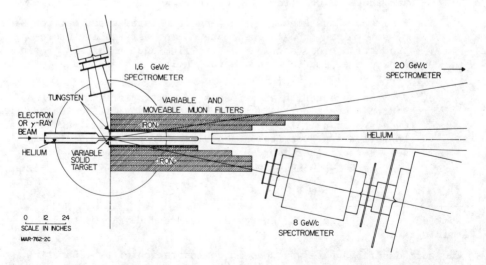

Fig. 1. Plan view of target area for single arm ψ and excess lepton experiment showing details of muon filters.

The 1.6 GeV/c spectrometer was set up to view the target under fixed conditions and provided a stable relative monitor.

Between the target and either spectrometer was a variable thick iron muon filter, which could be adjusted to stop all particles except muons from entering the spectrometers. The filters were suspended from a frame which could rotate about the target center and translate radially from the target center in the horizontal plane. The filters were composed of 3 or 6 separate iron slabs (for the 20 GeV/c or 8 GeV/c spectrometers respectively), which were independently suspended from the frames and remotely retractable in any combination so that they could be completely raised above the flight paths of the relevant particles. The slabs nearest the target were faced with tungsten to increase their stopping power. The number of muons produced as a result of pion decays between the target and the iron could thus be varied by changing the flight path between the target and the filter iron, either by moving the iron sideways or by retracting the front slabs. The effective thickness of the iron between the target and spectrometer could be controlled by varying the number of slabs retracted and by varying the angle of the filters relative to the spectrometer. In one mode of operation, the filters were rotationally locked to the spectrometers so that the relative angle could not change as the spectrometer angle was varied. The slab combinations, positions, and angles were recorded by the computer.

The on-line computer, an SDS9300, recorded on tape the settings and status of the equipment, event-by-event data, scaler readings, and beam monitors. It also provided control of the spectrometers, monitored rates, analyzed samples of the data and created paper and microfilm records of the data. Subsets of the data were transferred to the SLAC computation facility for additional analysis as the experiment progressed.

ENERGY-DEPENDENCE OF THE CROSS SECTION

To measure the production of ψ's, we made sweeps in $p_\perp$ with the 8 GeV spectrometer at 10 GeV, 12 GeV, 14 GeV, and 16 GeV maximum bremsstrahlung energies. The detected momentum at the target was -5.5 GeV/c for a muon. $p_\perp$ was varied by varying the detection angle. Both the pion and the muon spectra were measured in separate sweeps with and without a hadron filter.

Figure 2 shows the muon yields versus $p_\perp$. The pion spectra were fitted to the form: yield (pions) = $A_\pi \exp(-B_\pi p_\perp)$ (with $B_\pi \sim 6$). The resulting spectrum for muon decay should be of the form:

$$\frac{A_\mu e^{-B_\pi p_\perp}}{B_\pi p_\perp} .$$

The calculated Bethe-Heitler[2] directly produced muons have an almost identical shape. We, therefore, fitted our muon yields to the three parameter form:

$$Y_\mu = \frac{A\,e^{-Bp_\perp}}{Bp_\perp} + C(\psi\text{-shape}) \; .$$

The first term represents muons from pion decay and B. H. production. The values of A and B obtained were close to those that were obtained by calculation from the pion spectrum and B. H. production.

The ψ-shape was obtained from a Monte Carlo calculation. The hard lines in Fig. 2 are computer fits to the yield curves. The dotted lines show the extrapolation of the background terms.

From the values of C obtained, the cross section for ψ's was unfolded. Figure 3 shows the result of $|d\sigma/dt|_{t_{min}}$ evaluated at t_{min}. The major source of error is the value b of the slope parameter to be used for this determination. We used a value of $b = 2 \text{ GeV/c}^{-2}$, which could be in error by a factor 1.5.

We conclude on the basis of Fig. 3 that the new measurements, the Cornell measurements, and our old SLAC measurements are all in reasonable agreement and show a substantial increase at 12 GeV. This is in line with Vector Dominance models that would expect a substantial increase in the ψ-nucleon cross section at the energy for which the charmed D particles can be produced.

ψ-NUCLEON CROSS SECTION VIA A-DEPENDENCE

To measure the ψ-nucleon cross section directly, photoproduction of ψ's was measured from a Beryllium and a Tantalum target, each of 0.3 radiation length thickness. The ψ-nucleon cross section is determined from the measured ratio of the cross sections.

Measurements were made on single muons at a $p_\perp$ of 1.65 GeV/c with the SLAC 20 GeV spectrometer. The primary peak bremsstrahlung energy was 20 GeV and the detected muons had a momentum of -9 GeV/c at the target.

Empty target subtractions even for Tantalum were less than 5%. About 1100 events

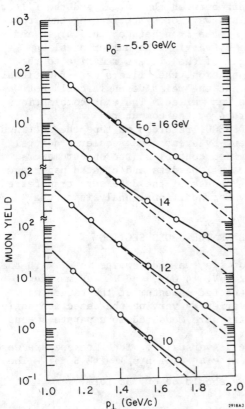

Fig. 2. Muon yields at incident photon energies of 10, 12, 14 and 16 GeV/c for a muon momentum at the target of -5.5 GeV.

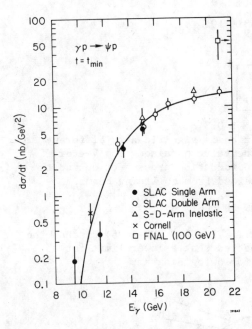

Fig. 3. dσ/dt at t = t_{min} for photoproduction. Black circles are this experiment. Note the FNAL point is off to the right on the graph.

were observed from the Tantalum target and about 4000 events from the Beryllium target. To evaluate backgrounds, the pion fluxes were measured from the respective targets and a $p_\perp$ sweep was made for the muon flux to determine the backgrounds from sources other than ψ-decays. Figure 4 shows the detected muons fitted for direct Bethe-Heitler production and ψ-decay. At a $p_\perp$ of 1.65 GeV, 11% arise from pion decay, 20% arise from Bethe-Heitler production, and 69% are from ψ-decay.

The raw measured ratio $|\mu(Be)/\mu(ta)|_{raw}$ = 1.19 ± .04. The measured ratio of pions from the targets was 1.18 ± .02 and hence the ratio for background muons arising from pion decay should also be 1.18 ± .07. The calculated ratio for Bethe-Heitler production is 1.03 ± .02.

If the raw ratio is corrected for the background, the corrected ratio $|\mu(Be)/\mu(Ta)|_{\psi's}$ = 1.25 ± .055. Nuclear physics corrections to the ratio are 6% arising from Pauli suppression, 3% from coherent production and -1% from the Fermi-momentum corrections to cross section for ψ-production. The ratio corrected for nuclear physics is then

$$|\mu Be/\mu Ta|_\psi^{Corr} = 1.17 \pm .055 \ .$$

To determine the ψ-nucleon cross section from this result, we used the standard model with $A_{eff}/A = 1-\delta/A$

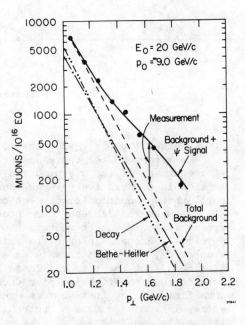

Fig. 4. Muon yield curve obtained with 20 GeV/c spectrometer from a 4-inch Beryllium target used to determine the A-dependence of ψ photoproduction.

$$\frac{\delta}{A} = \frac{9}{16\pi}(\frac{\sigma_{\psi N}}{r_o^2})A^{1/3} = 1.33 \cdot 10^{-2} \sigma_{\psi N} A^{1/3} \text{ (mb)}$$

yielding

$$\boxed{\sigma_{\psi N} = 2.75 \pm 0.90 \text{ mb} \\ \text{statistical error only.}}$$

The estimated systematic error is ~ ± 0.5 mb.

Therefore ψ-nucleon cross section is significantly different from zero mbs. Our result agrees with the value obtained from Vector Dominance arguments using the photoproduction cross section and the value of the photon-ψ coupling obtained from colliding beams.

EXCESS LEPTON PRODUCTION

Sivers, Townsend, and West (S.T.W.) (SLAC PUB-1636) discuss the photoproduction of D's. It is expected that a lower limit for the production of D's is given by the following diagram:

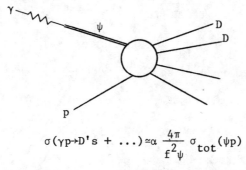

$$\sigma(\gamma p \to D's + \ldots) \simeq \alpha \frac{4\pi}{f_\psi^2} \sigma_{tot}(\psi p)$$

$$\simeq 500 \text{ nanobarns}$$

S.T.W. give $\sigma(\gamma p \to D's) \geq 300$ nbarns.

Therefore unlike the case for hadron production there exists a firm estimate of the D's produced via photoproduction. The cross section is comparatively large, being of the order of .4% of the total photoproduction cross section.

We measured "excess" muon production from a Beryllium target to see if we could detect muons produced from non-conventional (new physics) sources. In order to eliminate the muons from pion and kaon decay, measurements were made with the hadron filter at various distances from the target. These displacement curves were extrapolated to zero effective pion decay length to determine the prompt muon flux arising from sources other than decays. To make these extrapolations, we made a Monte Carlo calculation to take into account the multiple scattering effects on the input spectrum and the aperture of the spectrometer. Lateral displacement curves (Fig. 5) were taken at end point energies of 8 GeV, 12 GeV, and 20 GeV and at a $p_\perp$ of 1 GeV^2. The slopes of the lateral displacement curves were used to calibrate the effective aperture of the spectrometer so that prompt

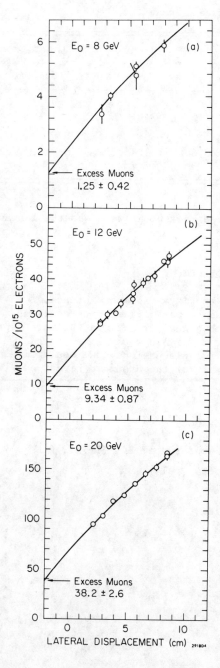

Fig. 5. Excess leptons at 8, 12, and 20 GeV photon energies obtained by displacing muon filters relative to the target.

muons arising from Bethe-Heitler production could be calculated.

Results are given in Table I. The signal at 8 GeV and 12 GeV is accountable from the prompt muons arising from Bethe-Heitler pair production. There appears to be a significant signal at 20 GeV with a μ/π ratio ~ 10^{-4}. This is similar to that observed in hadron experiments. To eliminate systematic effects, the only variable changed was the energy of the primary bremsstrahlung beam. Therefore the secondary momentum of the muons was held constant. A variation of primary energy, therefore, also corresponded to a variation in Feynman x. It is, therefore, possible to attribute the observed variation in the μ/π ratio to either the variation in Feynman x or to the variation with energy.

If all the excess muons at 20 GeV are attributed to charmed D decays, we can compare our result with model estimates. If we assume D's are produced with an integrated cross section proportional to the ψ-cross section, and with a shape

$$\frac{d\sigma}{dp^2 dx} \alpha\ e^{-5m_\perp}(1-x)^2$$

(of course, other assumptions can be made)

$$m_\perp = \sqrt{m_D^2 + p^2}$$

then our result is $\sigma(\gamma p \to D)$ x(Branching ratio to D's) = 2 10^{-33} cm^2 at a $p_\perp$ of 1 GeV/c and a muon momentum of ~5.5 GeV/c.

Estimating from theory, by combining the S.T.W. cross section with a 5% branching ratio gives $\sigma(\gamma p \to D)$ x BR ≥ 15 10^{-33} cm^2, a factor seven times higher than the observed value. This disagreement is uncomfortable, but it cannot be ruled out that other production

TABLE I. Excess Muons

E (GeV)	Total Excess/10^{15} e's	Calc B.H.	Net Excess (Total+B.H.)	Net Excess In Units Of $\mu/\pi \cdot 10^5$
8	1.25 ± 0.42	1.15 ± 0.25	0.1 ± 0.5	1.4 ± 7.0
12	9.34 ± 0.87	6.90 ± 1.70	2.4 ± 2.0	4.8 ± 4.0
20	38.20 ± 2.60	14.40 ± 3.60	23.8 ± 4.4	14.0 ± 2.5

spectrum assumptions would produce agreement between observation and expectation.

ACKNOWLEDGEMENT

We would like to acknowledge the invaluable assistance of Paul Tsai in the calculation of the Bethe-Heitler cross sections. We have used his computer program with a nuclear form factor more appropriate for large momentum transfer. We also would like to thank Fred Gilman and Stanley Brodsky for useful discussions. We would like to thank W. K. H. Panofsky and the staff of the Stanford Linear Accelerator Center for their support.

REFERENCES

1. cf. R. Prepost, Lepton Photon Conference, 241 (1975).
2. Y. Tsai, Rev. Mod. Phys., 46, 815 (1974).

SEARCH FOR NEW STATES WHICH DECAY SEMI-LEPTONICALLY*

SOD COLLABORATION

V. Cook, S. Csorna, D. Holmgren, A. M. Jonckheere,
H. J. Lubatti, K. Moriyasu, B. Robinson, P. Malecki,
Department of Physics, University of Washington,
Seattle, Washington 98195; D. Fournier, P. Heusse,
J. J. Veillet, Laboratoire de l'Accelerateur Lineaire,
Orsay, France; J. H. Klems, W. Ko, R. L. Lander,
D. E. Pellett, Department of Physics, University of
California, Davis, California 95616

Presented by H. J. Lubatti
Visual Techniques Laboratory, Department of Physics
University of Washington, Seattle, Washington 98195

The results of a search for new particle production with decay modes which include neutral strange particles (V^o) and muons are reported.[1] The number of neutral strange particles produced by interactions of a pion beam on a nuclear target was measured when muons were and were not required in the final states. Excess production of V^o's associated with a muon could result from production of a pair of new states ($D\bar{D}$) which can decay:[2] (i) both semi-leptonically; (ii) one leptonically and the other semi-leptonically, (iii) both leptonically. Cases (i) and (ii) would give an increase in the V^o rate observed in the streamer chamber and the momentum distribution of these V^o's would reflect their origin. The cross section with which we would observe this phenomenon is given by

$$\Sigma \equiv \left[B_{SL}^2 g_1 + 2 B_{SL} B_L g_2 \right] \sigma_{+-} \qquad (1)$$

where B_{SL} and B_L are the branching ratios of the new state into leptonic and semi-leptonic modes, the g_i are the appropriate efficiencies for detecting 2 muons and at least one V^o, and σ_{+-} is the cross section for producing charged particle anti-particle pairs.

A beam of negative pions of 225 GeV was incident on a Lucite ($C_5H_8O_2$) target, 5 cm thick, at the entrance of the University of Washington streamer chamber[3] (see Fig. 1). The apparatus shown in Fig. 1 consists of a 1 x .5 x .3 m³ streamer chamber placed in a large magnet whose field in the visible volume is approximately 8 kG. Immediately downstream of the streamer chamber is a muon filter which consists of lead and iron absorbers. A 3m magnetized iron absorber (SOD magnet) of cross section 0.56 m x 0.9 m transverse of the beam with $\int Bd\ell \simeq 56$ kG-m hardens the muon spectrum and determines the

---

* Work supported in part by ERDA, NSF, and the French CNRS.

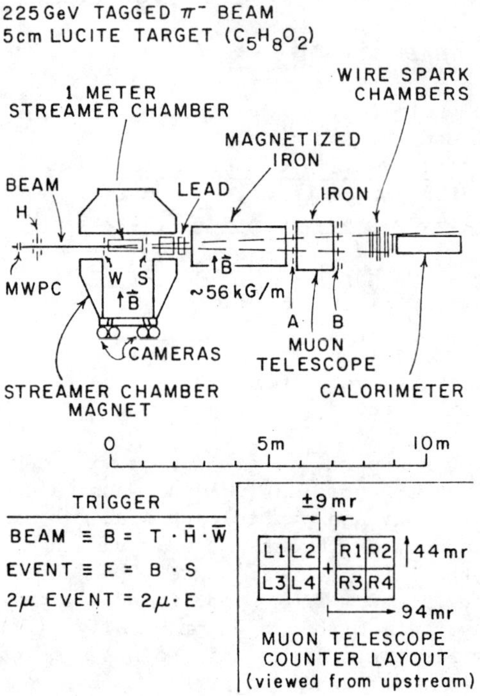

225 GeV TAGGED π^- BEAM
5 cm LUCITE TARGET ($C_5H_8O_2$)

Fig. 1. Experimental Layout

muon momentum. Both the streamer chamber and SOD magnets bend in the vertical plane. The muon telescope consists of two planes of 8 scintillation counters each separated by 48 inches of iron. The acceptance is ± 94 mr in the horizontal plane and ± 44 mr in the vertical plane. A gap of ± 9 milliradians is left in the central plane (see inset Fig. 1). A small acceptance calorimeter, 12" x 12", was used during part of the run in order to determine the hadronic punch through. This was found to be negligible.

The beam is defined by a beam telescope (T) and hole veto counters (H,W) to be $B \equiv T \cdot \bar{H} \cdot \bar{W}$. A 5 cm lucite target $C_5H_8O_2$ is placed at the upstream end of the streamer chamber. A scintillator (S) which has a 4.5 cm diameter hole for beam exit defines the interaction. Thus, an event trigger is $E \equiv B \cdot S$. A muon is defined to be a coincidence between two corresponding scintillators in the A and B planes (eg, $LA_i \cdot LB_i$). During the experiment, the beam intensity was in the range $2-7 \times 10^5$ π^-/pulse with an 800 ms spill.

Downstream of the muon telescope, three planes of wire spark chambers provide muon bend angle information; by swimming the tracks backward to intersect in the target at the upstream end of the streamer chamber, the charge and momentum of the muon is obtained.

The 1μ acceptance as a function of Feynman x and P_T is given in Fig. 2a.[4] We note that for P_T < 600 MeV/c, the acceptance falls rapidly. The k^0 detection efficiency in the streamer chamber is given in Fig. 2b.

Trigger rates for the E, 1$\mu \cdot$E and 2$\mu \cdot$E triggers are given in Table I along with the corresponding cross sections. In order to avoid scanning biases, these triggers were mixed. From the 1μ rate, we find $(1/2)(1\mu \cdot E/E)^2 \sim 0.3 \times 10^{-6}$, which is an upper limit on the 2μ rate we expect from pion or kaon decay. The factor of 1/2 compensates for double counting. Thus, we would expect the 2μ rate from π

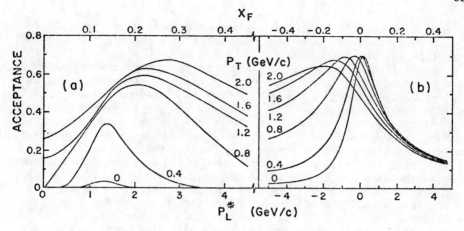

Fig. 2. Detection efficiency a) single muon b) k^0_s

Table I Trigger Rates and Cross Sections

Trigger	Rates	σ(A 2/3)	Scanned Events
E	E/B = 0.036	19.6 mb	1910
$1\mu \cdot E$	$1\mu \cdot E/E = 0.8 \times 10^{-3}$	15.7 µb	–
$2\mu \cdot E$	$2\mu \cdot E/E = 4.3 \times 10^{-6}$	84.3 nb	837

and k decay to be at least an order of magnitude less than the 2μ rate that is actually observed.

An independent check of the contamination has been made by taking $\bar{\pi}$p interactions at 200 GeV in the 30" FNAL bubble chamber from the Berkeley-FERMILAB collaboration.[5] We have calculated for each event and each charged prong, the probability of a π or K decay resulting in a trigger and find a rate which is 0.02 of our observed 2μ rate. Furthermore, we find that the like-charged dimuons are $\sim$ 5% of the oppositely-charged muons. Thus, we conclude that we are triggering on prompt muons.

A more complicated issue is whether the muons are, in fact, being produced in the target or in the absorber. We have checked this in two ways:[6] i) by taking those events in which two tracks can be reconstructed in the wire spark chambers, projecting them back through the absorber, and determining their intersection point in the horizontal (non-bend) plane; ii) by comparing χ^2 for fits to the wire chamber data which require that the track originate in the target and one interaction length into the absorber.[7] The two methods give results in good agreement, and we conclude that 50 ± 5% of our 2μ·E triggers are prompt muons from interactions in the 5 cm target in the streamer chamber.

A detailed scan of the 2μ·E and E triggers was made and all V^o's were recorded and measured. Since the streamer chamber contains Ne gas, there is a very low probability that an observed V^o is a conversion electron-positron pair. However, γ's can convert in the central wire plane. In order to insure the purity of the V^o sample, we have removed those V^o's whose vertex occurs within ± 1.5 cm of the central plane. (This removes approximately 25% of the sample.)

We find that the fraction of events which contains V^o's denoted by α is the same for the 2μ·E and the E triggers:
α_E = 0.102 ± 0.006 and $\alpha_{2\mu \cdot E}$ = 0.097 ± 0.010. Thus, we obtain an upper limit with 90% confidence for expression (1) of Σ < 1 nb. In calculating this upper limit, we assume that the two samples of data are identical and that the contamination of 2μ·E triggers by muons produced in the absorber gives events in the streamer chamber which are identical to the events in the E trigger.

From Fig. 3, we observe that the transverse momentum distribution of the neutral strange particles and the average number of V^o's produced as a function of charged multiplicity is within statistics identical for the two triggers. The multiplicity distributions of the charged particles produced are also the same for the two triggers (c.f. < n > and D in Table II).[8] There are no significant differences between events obtained with the 2μ·E trigger and the E trigger. Our conclusion is that we do not produce neutral strange particles which are correlated with muons at an observable cross section greater

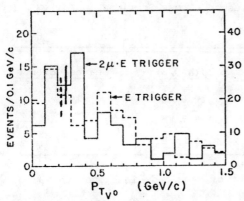

Fig. 3a. Transverse momentum distributions of V^o's. Left and right vertical scales apply respectively to the 2μ·E and E triggers.

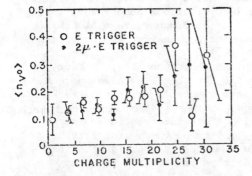

Fig. 3b. Average number of V^o's $<n_{V^o}>$ vrs. charged multiplicity.

Table II Average Charged Multiplicity and Dispersion of the 2μ·E and E Triggers

Trigger	n_{ch}	$D = [<n_{ch}^2> - <n_{ch}>^2]^{\frac{1}{2}}$
All E	12.5 ± 0.3	6.2 ± 0.3
All 2μ·E	12.3 ± 0.5	6.5 ± 0.3
V^0(E)	13.3 ± 0.9	6.6 ± 0.7
V^0(2μ·E)	13.5 ± 1.4	6.7 ± 1.1

than 1 nb in our 2μ·E trigger.

This result can be used to place limits on charmed particle production. For instance, GIM^2 D mesons of mass 3 GeV (5 GeV) each with exp($-P_T^2$) and flat x_F distributions observed in ψ production, yields an upper limit of $\sigma_{+-} B^2 < 35$ (25) nb, assuming $B_{SL} = B_L \equiv B$ in relation (1).

We wish to thank R. R. Wilson for making the SOD magnet possible, P. Koehler and the staff of the Meson Laboratory at FNAL for assistance and encouragement, and the scanning staffs of our respective collaborators for their careful work. We are indebted to R. Kenyon, D. Forbush, P. Rancon, and G. Wurden for technical assistance and C. Jones for preparing the manuscript.

REFERENCES

1. Preliminary results have already been reported: A. Jonckheere, et al., Bull. Am. Phys. Soc. 21, 36 (1976).
2. S. L. Glashow, J. Iliopaulos, and L. Maiani, Phys. Rev. D2, 1285 (1970); M. K. Gaillard, B. W. Lee, and J. L. Rosner, Rev. Modern Physics 47, 277 (1975).
3. V. Cook, et al. Proceedings of the International Conference on Instrumentation for High Energy Physics, Frascati, 171 (1973).
4. Due to the ± 9 mr gap in the muon telescope (see Fig. 1).
5. D. Bogert, et al., Phys. Rev. Lett. 31, 1271 (1973).
6. Spark chamber data was taken with a 2μ·E trigger which allows us to study muons from interactions in the absorber.
7. The χ^2 takes into account the measurement error and multiple scattering in the absorber.
8. $<n_{ch}>$ is larger than observed in $\pi^- p$ interactions because many of the 2 and 3 prong events go through the hole in the s counter and do not trigger. Also, secondary interactions in the nucleus and the 5 cm target increase $<n>$.
9. K. J. Anderson, et al., Phys. Rev. Lett. 36, 237 (1976).

STUDY OF PROMPT MUONS AND ASSOCIATED PARTICLES
IN n-Be COLLISIONS

R.C. Ruchti
Northwestern University, Evanston, Il. 60201

ABSTRACT

We are currently studying the associated production of prompt muons and other particles produced in n-Be collisions in the M-3 neutral beam at Fermilab. The muon arm of our two-arm spectrometer provides excellent hadron rejection and accepts particles of transverse momentum greater than 0.35 GeV/c. The muon momentum is measured with an accuracy of about ±25%. The precision forward arm of the spectrometer has large acceptance (x > 0.2) and is instrumented with equipment for lepton identification. The system is triggered whenever one or more charged particles traverse the muon arm in conjunction with at least one charged particle in the forward arm. Preliminary distributions for particles in the muon arm will be presented.

We present a progress report of an experiment[1] to search for final states produced in association with a prompt muon via the process:

$$n + Be \rightarrow \mu \text{ (prompt)} + \text{anything}.$$

The detector is situated in the M-3 neutral beam at Fermilab which delivers neutrons of ~300 GeV/c average momentum onto a 1/10 interaction length beryllium target. The spectrometer (fig.1) consists of an upward arm to define the muon and measure its charge and momentum ($35 \leq \theta \leq 100$ mr vertical acceptance, and ±90 mr horizontal acceptance), and a high resolution, large-acceptance forward arm to detect the associated final state (±16 mr vertical acceptance, and ±60 mr horizontal acceptance). Lepton identification in the forward arm is provided by a 38 element electron shower counter array (e) and a 12 counter muon hodoscope (μ), permitting μe, $\mu\mu$ correlations and μ + multihadron final states to be studied simultaneously. In particular we are sensitive to the associated production of hypothesized Charmed Particles, for example:

$$n + Be \rightarrow D^0 + \overline{D}^0 + \text{anything}$$
$$\hookrightarrow \mu + \text{anything}$$
$$\hookrightarrow \mu, e, K\pi, K\pi\pi + \text{anything}$$

where the muon from the semi-leptonic decay of the $\overline{D}^0$ traverses the muon arm, and the semi-leptonic or weak hadronic decay of the associated D^0 is observed in the forward arm.

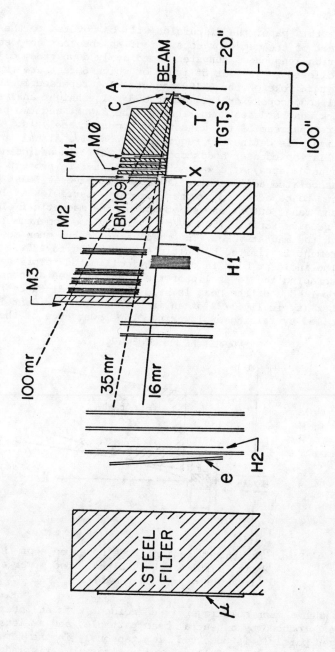

Fig. 1 Elevation view of the apparatus.

In this paper the discussion will be devoted to the muon arm. This part of the detector consists of an absorber composed of two absorption lengths of tungsten immediately downstream of the target, followed by 16 feet of iron, of which 6 feet are the polarized return yoke of the BM109 magnet. Interspersed in the iron are several hodoscopes for triggering and pulse height analysis and 16 spark chamber gaps downstream of the magnet and two proportional chambers upstream of the magnet for momentum analysis. A muon sub-trigger is defined by requiring a positive signal in each of the scintillation counter hodoscopes M0,M1,M2,M3 in anticoincidence with counter array A, a wall of scintillator to protect against beam associated accidentals. A muon must have at least 6 GeV/c momentum to be accepted by the detector - particles of lower momentum range out or are swept out by the magnetic field. The muons receive roughly 750 MeV/c of transverse momentum in passing through the magnet yoke. The resolution of the muon momentum measurement is limited by multiple scattering to ±25%.

The incident beam is carefully collimated to prevent halo from scraping the lower lip of the iron-tungsten absorber (fig.2). The upper edge of the beam is held at a fixed distance from the absorber (typically 1/32 of an inch), and the incident intensity is controlled by raising or lowering the lower edge of the beam.

TARGET REGION (SCHEMATIC)

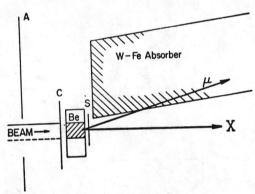

Fig.2 The target region. The upper edge of the beam (solid line) is fixed. The lower edge (dashed line) is adjustable.

When a muon subtrigger is coincident with an interaction in the target produced by a neutral beam particle, and at least one charged particle is observed simultaneously in the forward arm, ($\overline{C} \cdot S \cdot H1$), then the full two-arm trigger is satisfied. In 70% of the events there is a clean muon track visible in the spark chambers downstream of the BM109 magnet. Approximately 80,000 triggers have been recorded with incident neutrons of ∿300 GeV/c average momentum.

The distributions of momentum and transverse momentum for particles in the muon arm when the full trigger is satisfied are shown in figures 3 and 4. The acceptance of the apparatus is essentially flat for muon momenta greater than 10 GeV/c and transverse momenta greater than 0.4 GeV/c but falls off dramatically to zero below these values.

To determine what fraction of the muon candidates are prompt, we have made several tests: we have varied the vertical beam-spot size and have removed material from the absorber near the target. How the data rates were affected by these tests are shown in

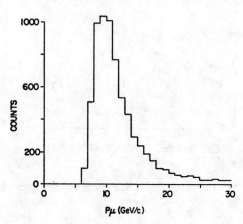

Fig. 3 Muon momentum distribution, uncorrected for acceptance.

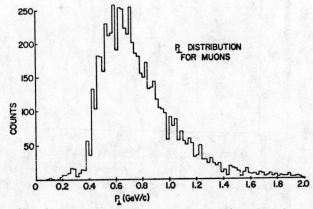

Fig. 4 Muon transverse momentum distribution, uncorrected for acceptance.

figures 5 and 6 and indicate a 9 inch decay length. If prompt muons occur at roughly 10^{-4} of the pion yield, then we have a signal containing approximately 20% prompt muons. In addition we observe dimuon events contained within the muon arm at a rate which is of order 2% of that for single muons. This is consistent with preliminary estimates for the production of vector mesons within our acceptance, and strengthens our conviction that we are indeed observing a prompt signal.

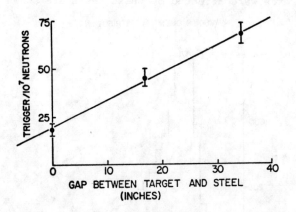

Fig. 5 Trigger rate as a function of gap size between the target and the absorber.

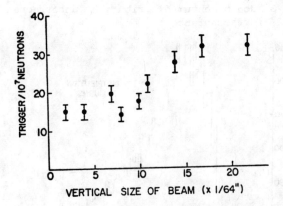

Fig. 6 Trigger rate as a function of beam spot size.

The data (fig.3) indicate that we are sensitive to prompt muons in the very interesting range $0.4 \leq P_\perp \leq 1.0$ GeV/c. Recent results from the ISR[2] indicate that the lepton to pion yield may be large in this domain. Once our event analysis and acceptance studies are complete, the values for the $\mu^\pm/\pi^\pm$ ratios will be made available.

I would like to thank my collaborators in the Northwestern-Rochester-Fermilab group for helpful discussions.

REFERENCES

1. Fermilab Experiment E-397, Northwestern-Rochester-Fermilab collaboration.
2. C.Rubbia, paper presented at this conference.

DIMUON PRODUCTION BY PIONS AND PROTONS IN IRON
AND A SEARCH FOR THE PRODUCTION IN HYDROGEN OF
NEW PARTICLES WHICH DECAY INTO MUONS*†

Presented by M. L. Mallary for:

G. J. Blanar[□], C. F. Boyer, W. L. Faissler, D. A. Garelick[‡],
M. W. Gettner, M. J. Glaubman, J. R. Johnson, H. Johnstad[δ],
M. L. Mallary, E. L. Pothier, D. M. Potter, M. T. Ronan,
M. F. Tautz, E. von Goeler and Roy Weinstein
Northeastern University at Boston, Boston, MA 02115

ABSTRACT

Dimuon production cross sections have been measured with a 200 GeV π^- beam and a 240 GeV proton (p) beam. The dimuon mass spectra produced by π and p have essentially the same shape. The dimuon transverse momentum behavior is essentially the same for vector mesons produced by π or by p. For ($\rho+\omega$) or ψ production the x (Feynman) behavior is approximately independent of the mass of the dimuon produced. The dimuon π production cross section is approximately 6 times greater than that of the p for x > .5.

Using a missing mass (M_m) technique, the same pion and proton beams were used to search for the production in hydrogen of new particles which decay with muon emission. Model dependent upper limits (50 to 7000 nanobarns) for the production of such objects with $2.5 < M_m < 7.$ GeV are given (for charmed meson pair production M_m is the mass of the $D\bar{D}$ system).

INTRODUCTION

This talk is a report on the results from two experiments for which the data were collected simultaneously in the M2 beam line of the Meson Laboratory at Fermi National Laboratory. Data were taken with a negative pion beam at a momentum of 200 GeV and a proton beam at a momentum of 240 GeV on the following reactions:

$$\pi^- + Fe \rightarrow \mu^+ + \mu^- + \text{anything} \quad (1)$$

$$p + Fe \rightarrow \mu^+ + \mu^- + \text{anything} \quad (2)$$

$$\pi^- + p \rightarrow p_{recoil} + Y \searrow_\mu + \text{anything} \quad (3)$$

$$p + p \rightarrow p_{recoil} + Y \searrow_\mu + \text{anything} \quad (4)$$

* Work supported in part by the National Science Foundation under Grant No. MPS70-02059A5.
† Data taken at the Fermi National Accelerator Laboratory.
□ Address: CERN, Geneva, Switzerland.
‡ Supported in part by an Alfred P. Sloan Foundation Fellowship.
δ Address: Faculte des Sciences, Univ. de l'Etat, Mons, Belgium.

In the data from reactions (1) and (2) the dimuon invariant mass ($M_{\mu\mu}$) spectra were examined for structures from the production of η, ρ, ω, φ, and ψ mesons. For both the $\rho + \omega$ and the ψ, the dependence of the data on longitudinal and transverse momenta were fit to phenomenological forms.

For reactions (3) and (4) the missing mass (M_m) spectra of the system (Y) recoiling against the recoil proton were examined for threshold enhancements in the mass range $2.5 < M_m < 7.$ GeV. Such an enhancement would be evidence for the production of new particles which decay into muons such as charmed particle pairs, vector bosons or heavy lepton pairs. Examples of low mass threshold enhancements have been seen previously for familiar reactions.[1]

Apparatus

The apparatus for these two experiments is shown in Figure 1. It consists of a hydrogen target, a recoil proton spectrometer, and a forward muon spectrometer. Beam optics and counters are not shown. The proton spectrometer[2] measured the recoil proton angles with four magnetostrictive wire spark chambers (SC1 - 4). The proton energy was measured by pulse height analysis of the E counters, in which the protons stopped. The pulse heights from the dE counters were used to identify the stopping particles as protons. The A counters vetoed events in which the protons did not stop. The t acceptance of this spectrometer was $.1 < -t < .4$ GeV^2. The axis of the spectrometer was set at 70° to the beam giving good acceptance in the missing mass interval $0 < M_m < 7$ GeV. The M_m^2 resolution was 2 GeV^2 FWHM.

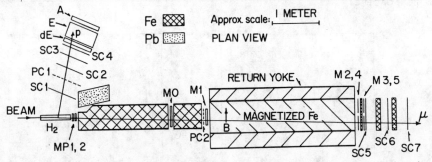

Fig. 1 Plan view of the apparatus. There are five dE, E, and A counters stacked vertically.

The forward muon spectrometer[3] consisted of a 2 m long iron hadron absorber followed by a 56 kilogauss-m magnetized iron spectrometer which bent the muons vertically. Track coordinates were measured in front of the magnet by a multiwire proportional chamber (PC2) and at the rear of the magnet by three magnetostrictive wire spark chambers (SC5 - 7). The trigger for the muon spectrometer required pulses in counters M0·M1·(M2·M3 or M4·M5) for reactions (3) and (4) (at least one muon). For reactions (1) and (2), 1½ minimum ionizing pulse heights were required in M0 and M1 and all of the counters M2·M3·M4·M5 were required (at least two muons). The pulse heights in counters MP1 and MP2 were used off line to determine the forward charge multiplicity of events from reactions (3) and (4).

In the analysis of the data the muon trajectories were tracked back from the spark chambers to the production point. The momentum was determined from the bend in the vertical plane. In the horizontal plane the tracks were required to extrapolate to within 2.2 s.d. of the production point (errors based on multiple scattering at the measured momentum). In the analysis of the data from reactions (1) and (2) the production point was assumed to be on the beam line at a depth of one absorption length (17 cm) into the iron hadron absorber. For reactions (3) and (4) the production point was assumed to be the beam particle-recoil proton vertex inside the hydrogen target. The muon longitudinal momentum ($P_{\mu\ell}$) was reconstructed to an rms accuracy of approximately $\pm$ 16% and the transverse momenta ($P_{\mu\perp}$) to 0.2 GeV (for $P_{\mu\perp} \lesssim 1$ GeV). The apparatus accepted muons with $P_{\mu\ell} \gtrsim$ 20 GeV and $P_{\mu\perp}/P_{\mu\ell} \lesssim .05$.

Excellent agreement between $P_{\mu\ell}$ and $P_{\mu\perp}$ distribution for data taken on beam muons and the corresponding Monte Carlo distributions gives us confidence in our understanding of the apparatus and the Monte Carlo program.

Dimuon Analysis

In the following analysis the laboratory dimuon longitudinal momentum (P_ℓ) was required to be greater than 90 GeV. This reduces to a negligible level the production of dimuons from the interaction of secondaries in the iron hadron absorber. The dimuon invariant mass spectra from reactions (1) and (2) are shown in Figures 2a and 2b respectively.

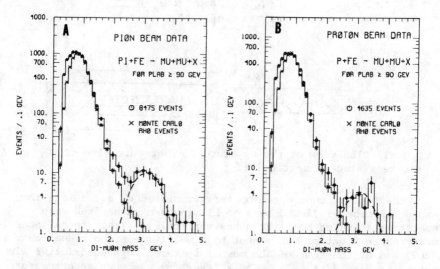

Fig. 2. Di-muon invariant mass spectra a) for reaction (1) and b) for reaction (2).

Also shown along with these spectra are Monte Carlo results (x's) which were generated with a relativistic s-wave Breit Wigner distribution in mass with the parameters of the ρ. The longitudinal and transverse momentum distributions of the Monte Carlo events were chosen to fit the data. The Monte Carlo event sample was normalized to the data in the mass range $1. < M_{\mu\mu} < 1.5$ GeV. From comparisons of the Monte Carlo simulated invariant mass distributions to the data, we conclude that 75% of the total observed dimuon production is consistent with the production of dimuons from the decay of ρ's and ω's (resolution doesn't allow a separation of ρ's from ω's).

In the low mass region ($M_{\mu\mu} \lesssim .6$ GeV) there is a sizable contribution (~ 25%) consistent with either Bethe Heitler production by γ secondaries or with $\eta \to \mu\mu\gamma$ decays. Monte Carlo studies also indicate that dimuon production from φ's is less than 15% of the ρ + ω dimuon production.[4] In the high mass part of these mass spectra ($2.5 < M_{\mu\mu} < 3.7$ GeV) there are enhancements from $\psi(3.1)$ decay[3]. The dashed curves shown with these spectra are gaussian distributions with widths of 13% rms (which is the ψ resolution expected from Monte Carlo calculations) centered at the ψ mass.

In the analysis of the longitudinal and transverse momentum spectra the events satisfied the longitudinal momentum cut, $P_\ell > 90$ GeV, and mass cuts of $2.5 < M_{\mu\mu} < 3.7$ for the ψ sample and $.77 < M_{\mu\mu} < 1.0$ GeV for the ρ + ω sample. The fits to the spectra were obtained by generating Monte Carlo simulated events with an isotropic center of mass decay distribution. Best fits were then found for phenomenological forms given in Eqs. 5, 6, and 7 below by adjusting the parameters in the Monte Carlo generations until the data was accurately simulated.

The longitudinal momentum distributions were fit to the following forms:

$$\frac{d\sigma}{dx} \alpha\ e^{-ax} \quad (5)$$

and

$$E_{\mu\mu} \frac{d\sigma}{dx} \alpha\ (1-|x|)^n \quad (6)$$

where a and n are parameters which were varied; $x = P_{\mu\mu}/P_{max}$; $E_{\mu\mu} = \sqrt{P_{\mu\mu}^2 + M_{\mu\mu}^2}$; $P_{\mu\mu}$ is the center of mass longitudinal dimuon momentum and $P_{max} \approx \sqrt{s}/2$. The best fits to these parameters are shown in Table I.[5] The measured cross section per iron nucleus times the muon branching ratio ($\sigma_{90}B$) for $P_\ell > 90$ is also shown in Table I along with the cross section per iron nucleus times the branching ratio ($\sigma_0 B$) for $x > 0$ obtained by extrapolating to $x = 0$ the fits with Equation (6) (preliminary results).

The spectra in dimuon transverse momentum ($P_\perp$) were fit to the form

$$\frac{d\sigma}{dP_\perp^2} \propto e^{-bP_\perp^2} \qquad (7)$$

The fits to this form for the $\rho + \omega$ data are shown in Table II.[4]

Table I Fits to the dimuon spectra in x (brackets indicate estimated systematic uncertainties)

Mesons	Beam	a	n	$\sigma_0 B$	$\sigma_{90} B$
				microbarns/Fe nucleus	
$\rho + \omega$	π	3.5 ± .2(.5)	0.8 ± .3(.5)	51 ± (26)	6.1 ± (3.)
$\rho + \omega$	p	8.8 ± .3(.5)	3.1 ± .1(.5)	45 ± (23)	2.5 ± (1.3)
ψ	π	5.7 ± .8	1.2 ± .2(.5)*	.6 ± (.3)*	.11 ± (.06)
ψ	p	9.0 ± 1.5	4.5 ± .5 *	.7 ± (.4)*	.04 ± (.02)

* Preliminary results from a larger data sample than shown in Figs. 2a and b.

Table II Fits to the dimuon spectra in $P_\perp^2$ (brackets indicate systematic uncertainties)

Meson	b(π Beam)	b(p Beam)	$P_\perp^2$ range (GeV^2)
$\rho + \omega$	2.6 ± .1(.2)	2.1 ± .1(.2)	0. - 2.
ψ	0.8 ± .2	1.1 ± .3	0. - 3.

From the fits to the dimuon spectra from Reactions (1) and (2) we find that the $P_\perp$ spectra depend strongly on the meson produced ($\rho + \omega$ or ψ) and weakly on the beam particle whereas the P_ℓ spectra depend strongly on the beam particle and weakly on the meson produced. For forward production x > .5, we find pions produce dimuons 5.8 ± .7 times more abundantly than protons for $\rho + \omega$ production and 7.4 ± 2.0 times more abundantly for ψ production. This reflects the fact that the pion and proton mass spectra have essentially the same shape.

New Particle Search

Reactions (3) and (4) were examined for threshold effects in the missing mass (M_m) spectra in association with muon production. Such an effect would be indicative of the production of new particles which decay into muons. The missing mass squared spectra for the data from Reactions (3) and (4) are shown in Figures 3a and 3b, respectively (corrected for proton acceptance but not for muon acceptance). For calibration purposes, data were also taken simultaneously without the muon requirement. The missing mass squared spectra of this data is shown in Figures 3c and 3d.

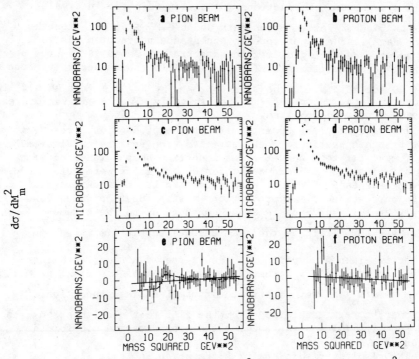

Fig. 3 a) and b) $d\sigma/dM_m^2 = \int (d\sigma/dt dM_m^2) dt$, $.1 < -t < .4$ GeV^2 for Reactions (3) and (4), respectively.

c) and d) $d\sigma/dM_m^2$ for the calibration data (no muon required).

e) and f) $d\sigma/dM_m^2$ difference spectra (see text).

The general features of these spectra are that they are similar in shape but that they differ in magnitude by a factor of about 1000 which is consistent with the expected inflight π and K decay probability. There is an elastic peak, a diffractive peak (not resolved from the elastic peak) which falls as an inverse power of the mass and a flat non-diffractive region.[6] In order to remove the diffractive structure from the spectra in Figure 3a and 3b, for the purpose of bump hunting, the raw mass spectra (not corrected for proton acceptance) of the calibration data (no muon) were fit to the phenomenological form:

$$dN/dM_m^2 = A + BM_m^2 + C/M_m^r \qquad (8)$$

where A, B, C, and r were free fit parameters. The functions which were obtained in this way from the pion and the proton calibration data were then normalized to and subtracted from the corresponding raw mass spectra from Reactions (3) and (4) (in the intervals $4 < M^2 < 54$ GeV^2 for the pion data and $6 < M^2 < 54$ GeV^2 for the proton data). The recoil proton acceptance correction was then made and the resulting difference spectra are shown in Figures 3e and 3f. Straight line fits (solid lines) to these spectra yielded a χ^2 of 49 for 48 d.f. in the pion data and a χ^2 of 45 for 46 d.f. in the proton data. There are no statistically significant structures in these difference spectra.

In order to place upper limits on the production cross section times branching ratio for new particles, the spectra in Figures 3e and 3f were fit to a number of assumed line shapes (smeared with resolution) and 2 standard deviation upper limits were determined. The line shapes used were: a relativistic Briet Wigner to place limits on resonance like production; a step function which rose sharply at threshold and remained constant up to the kinematic limit $M_m^2 \approx 2E_B \sqrt{-t}$ (E_B is the beam energy) to set limits on non-diffractive like production;[6,7] and a step function which rose sharply at threshold and then fell as a power of the mass ($d\sigma/dM_m^2 \sim 1/M_m^r$) to set limits on diffractive like production.[6,7] The power, r, was determined from the fits to the calibration data with Equation (8) and was 3.2 for the pion data and 4.0 for the proton data.

The results of this analysis are given in Table III. The entries are the 2 s.d. upper limits on the production cross section (σ_i) times the muon branching ratio (B_μ) times the muon acceptance (A_μ) integrated over the t interval $.1 < -t < .4$ GeV^2 and the mass squared interval $0 < M_m^2 < 2E_B \sqrt{-t}$.

Table III. Upper limits (2 s.d.) on $\sigma_i B_\mu A_\mu$ (in nanobarns).

Beam Particle	Approx. M_m^2 (GeV^2)	Resonance Prod. Γ=.05, .2, .8 GeV			Diffractive Production	Non Diffractive Production
π	7	31	43	85	90	3400
π	17	32	62	97	210	1400
π	36	37	54	116	510	1600
π	49	39	58	118	740	1800
p	10	73	92	161	190	3700
p	17	35	44	79	130	1900
p	38	42	57	115	430	2700
p	49	25	44	157	390	1900

To extract upper limits from this table on the total production cross section for massive particles which decay into muons, a specific hypothesis describing the production and decay of these particles is necessary. For example, if charmed meson pairs were produced diffractively[8] by pions, had a branching ratio to muons of unity, decayed to three bodies (e.g. $K\mu\nu$), were produced with a t distribution of $d\sigma/dt \sim e^{-5t}$, and had a mass of 2 GeV (pair mass squared of 16 GeV^2), the relevant entry in the table is 210 nanobarns. This limit is shown as a dashed curve on Fig. 3e. This number is then divided by two to correct for the fact that either meson of the pair can decay, and multiplied by 2 to account for the production outside of the t interval $.1 < -t < .4$ GeV^2. The muon acceptance must also be corrected for, and is then obtained from the following empirical formula:

$$A_\mu = 1 - (M_m/E_B)(5.6\sqrt{E_{cm\mu}} + 9/E_{cm\mu}) \qquad (9)$$

where $E_{cm\mu}$ is the muon center of mass energy in GeV in the forward Y system. For a 2 GeV charm mass and $K\mu\nu$ decay, $E_{cm\mu}$ is about .5 GeV so A_μ is about 0.6. Accounting for these effects, the 2 s.d. limit on diffractive charmed meson pair production, with the above assumptions, deduced from Table III, is 350 nanobarns.

REFERENCES

1. In $\pi + p \to$ nucleon + x, threshold enhancements are seen when x is: $\pi\rho$ by C. Caso et al., Nuovo Cimento, 54A, 983 (1968); πf by J. Bartach et al., Nuclear Physics, B7, 345 (1968); K^+K^- by B. D. Hyams et al., Nuclear Physics, B22, 189 (1970); and $\bar{p}p$ by G. Grayer et al., Phys. Lett., 39B, 563 (1972).
2. D. Bowen et al., Phys. Rev. Lett. 26, 1663 (1971).
3. G. J. Blanar et al., Phys. Rev. Lett. 35, 346 (1975); Proceedings of the 1975 SLAC Summer Institute on Particle Physics; Proceedings of the Division of Particles and Fields Meeting of the American Physical Society (August 1975) University of Washington, Seattle, Washington; R. Weinstein et al., Proceedings of the 1975 International Symposium on Lepton and Photon Interactions at High Energies, Stanford University.
4. This assumes that $\eta \to 2\mu$ is small. For an indication of this in pBe see the mass spectra from Yu. M. Antipov et al., Physics Letters, 60B, 309 (1976).
5. These results are described in detail by: M. Ronan, thesis, Northeastern University (1976); and reference (3).
6. F. C. Winkelmann et al., Phys. Rev. Lett. 32, 121 (1974).
7. M. Jacob et al., Phys. Rev. D6, 2444 (1972).
8. The diffractive line shape is a fair approximation to previously observed threshold enhancements (See reference 1).

SEARCH FOR NARROW HIGH MASS STATES AT FERMILAB*

M. A. Abolins, D. Cardimona, J. A. J. Matthews, R. A. Sidwell
Michigan State University
East Lansing, Michigan 48824

H. R. Barton, Jr., N. W. Reay, K. Reibel, N. Stanton
Ohio State University
Columbus, Ohio 43210

K. W. Edwards, G. Luxton
Carleton University
Ottawa, Ontario, Canada

(presented by J. A. J. Matthews)

ABSTRACT

A search has been made in neutron beryllium interactions for neutral narrow enhancements produced at Feynman $x \gtrsim 0.2$. Upper limits are presented for production cross sections times branching ratio into the 2-body channels $\pi^+\pi^-$, π^+K^-, pK^- and $p\bar{p}$ for masses in the interval 2 GeV $\lesssim m \lesssim$ 4 GeV.

INTRODUCTION

The discovery of the (J,ψ), ψ', ... mesons[1] suggests that a threshold has been crossed for the excitation of a new quantum number(s) within matter. We will refer to this generically as "charm". The charm conjecture then predicts a new spectroscopy of charmed mesons and baryons with typical masses $\gtrsim$2 GeV.[2] The lowest mass states will presumably decay weakly.

Our experiment has searched for the inclusive production of such long lived (charmed) states in neutron induced reactions. Two-body decay channels are analyzed for resonances produced forward in the center of mass, for example by associated production.

EXPERIMENTAL DETAILS

The experiment uses the forward spectrometer shown schematically in Fig. 1. The neutron beam is incident on a (typically) 0.1 interaction length beryllium target just upstream of the counter S_1. Charged secondaries are reconstructed by eleven proportional

---

*Work supported by the National Science Foundation, the U.S. Energy Research and Development Administration, and by the National Research Council of Canada.

chamber planes upstream and between 2-BM109 dipole magnets, and by fourteen planes of wire spark chambers downstream.

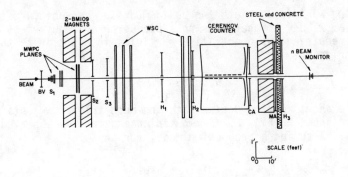

Fig. 1. Plan view of the spectrometer

Produced particles are analyzed by a segmented low pressure N_2 threshold Cerenkov counter instrumented with RCA 31000M phototubes. Pion threshold is set at 20 GeV/c. The Cerenkov is calibrated in place using 300 GeV/c diffracted protons down the neutral beam line. Muons are identified by a 10-foot range requirement in steel.

The neutron beam, produced at ~1 mr from 300 GeV/c pBe interactions in the meson laboratory target, peaks at ~240 GeV/c and has ≲1% K_L^o contamination above 100 GeV/c.[3] Photons in the neutral beam are removed using lead absorbers. The neutron flux is continuously monitored using a thin converter-counter telescope placed downstream of the spectrometer.

To obtain a mass selective trigger we utilize the technique of "point to parallel focusing" often used in K_s^o production experiments. Thus, the required transverse momentum "kick" from the dipole is approximately one half the 2-body mass. This geometry provides the optimal invariant mass resolution for a forward spectrometer, and the large magnetic field strongly defocuses low transverse momentum background tracks. The parallelism trigger requires ≥ 1 track on each side of the beam line plus "parallel logic" using hodoscope elements in H_1 and H_2. The finite mass bite of the spectrometer, Δm ~ 600 MeV, necessitates ~4 mass (magnet) settings to survey the mass interval ~2 GeV to ~4 GeV.

The spectrometer is sensitive to the production of high mass states at Feynman x ≳ 0.2, and with $p_\perp$ ≲ 1.0 GeV/c. The average laboratory momenta of the 2-body hadronic data is $<p_z>$ ~ 90-100 GeV/c corresponding to $<x>$ ~ 0.4. Thus, the typical particle momenta, ~45-50 GeV/c, fell below kaon Cerenkov threshold allowing pions to be separated from kaons and protons.

Running was divided into two basic modes: "hadronic", using the parallel 2-body trigger plus possible Cerenkov requirements, and "2μ", requiring two penetrating particles in addition to the parallel trigger.

RESULTS

From a short running period with the muon trigger we obtained the 2μ data shown in Fig. 2. A two bin enhancement is observed at the ψ mass. Assuming a production model for the ψ:

$$\frac{d\sigma}{dxdp_\perp^2} \approx f(x) \, e^{-ap_\perp^2}$$

and choosing $f(x)$ constant near $x = 0$:[4] (1)

$$f(x) = \begin{cases} 1.0 & \text{for } x < 0.35 \\ e^{-5(x-0.35)} & \text{for } x \geq 0.35 \end{cases}$$

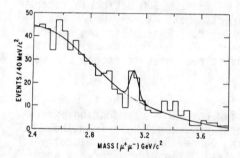

Fig. 2. $\mu^+\mu^-$ mass distribution

we obtained a ψ cross section times branching ratio into 2μ for $x \geq 0.2$ of 3.3 ± 1.8 nb/nucleon or 1.6 ± 0.9 nb/nucleon for $a = 1$ GeV^{-2} or $a = 3$ GeV^{-2} respectively. An "A" dependence of the cross section on atomic number was assumed. This cross section is consistent with (although greater than) the results of references 4 and 5.

The invariant mass distributions for the 2-body hadronic channels $\pi^+\pi^-$, π^+K^-, pK^- and $p\bar{p}$ are shown in Fig. 3. Data from the lowest mass (magnet) setting and the three higher mass settings are plotted separately; the former indicates the mass bite of the spectrometer. The relative normalization of these data is arbitrary. Additionally, some overlap does exist in these data since protons and kaons are indistinguishable below 70 GeV/c.

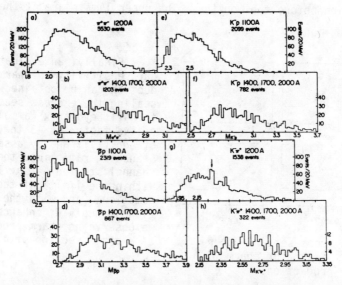

Fig. 3. 2-body hadronic mass distributions

These mass spectra are compared to smoothed distributions, for example generated with randomized tracks from different events, to search for ≥ 4 standard deviation enhancements. One such enhancement is observed in the π^+K^- mass distribution at $m_{\pi K}$=2.29±0.03 GeV, with a width consistent with our experimental mass resolution, see Fig. 4. Assuming the production model eqn. (1) with $a = 3$ GeV^{-2}, we obtain a cross section times branching ratio of 65 ± 26 nb/nucleon for $x \geq 0.2$. This corresponds to a production cross section of $\sim 40 \; \sigma_{\psi \to \mu\mu}$. We note that this "$4\sigma$" enhancement has a purely statistical probability of ~3%.

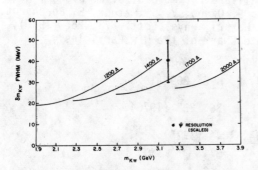

Fig. 4. $K\pi$ mass resolution at the 4 mass (magnet) settings

In the absence of enhancements, cross section times branching ratio upper limits for $x \geq 0.2$ are calculated corresponding to a "4σ" effect. These results, shown in Fig. 5, incorporate the experimental mass resolution and include corrections for secondary interactions and decays of particles, Cerenkov efficiencies, track reconstruction efficiencies and neutron beam attenuation before the monitor. The upper limits are at the level of $\gtrsim 20~\sigma_{\psi \to \mu\mu}$.

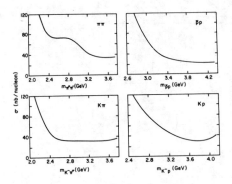

Fig. 5. Cross section upper limits for charm production.

Finally, an evaluation of our sensitivity to charm production can be obtained by scaling these cross sections for all Feynman x (a multiplicative factor of ~3 in the model of eqn.(1)), and by assuming a branching ratio into the 2-body channel studied -- say 10%. Within the limits of these assumptions (and with the exception of the π^+K^- enhancement), we observe no charm production at a cross section of $\gtrsim 1~\mu b$.

REFERENCES

1. J. J. Aubert et al., P.R.L. 33, 1404 (1974), J. E. Augustin et al., P.R.L. 33, 1406 (1974), G. S. Abrams et al., P.R.L. 33, 1453 (1974), and generally: Proceedings of "1975 Inter. Symposium on Lepton and Photon Interactions at High Energies", ed. W. T. Kirk (1975).

2. See for example: M. K. Gaillard, B. W. Lee, J. L. Rosner, Rev. Mod. Phys. 47, 277 (1975), S. Okubo, V. S. Mathur, S. Borchardt, P.R.L. 34, 236 (1975), H. Harari, Phys. Lett. 57B, 265 (1975), A. de Rujula, "The Discrete Charm of the New Particles", Harvard Univ., Preprint (1975).

3. P. V. R. Murthy et al., N. P. B92, 269 (1975).

4. K. J. Anderson et al., P.R.L. 36, 237 (1975).

5. B. Knapp et al., P.R.L. 34, 1044 (1975).

COMMENT ON SESSION CONCERNING NEW-PARTICLE SEARCHES[*]

T. Ferbel
University of Rochester, Rochester, N.Y. 14627

"Saying, Fight on, my merry men all,
And see that none of you be taine;
For I will stand by and bleed but awhile,
And then will I come and fight againe"

From Ionnë Armestrong

As chairman of this session I have been given the privilege of responding to the many excellent papers which have been presented. I will restrict my remarks to the properties of the J/ψ particle and to searches for charm-like objects in hadronic collisions.

Two oft-mentioned features of J production in hadronic collisions are: 1) the weak dependence of the J cross section on transverse momentum (p_T), and 2) the unusually rapid rise of the production cross section with increasing energy. The p_T dependence at Fermilab energies, for example, can be parameterized as $Ed\sigma/d^3p \sim \exp(-1.1 p_T^2)$. The average value of p_T is therefore quite large ($<p_T> \sim 0.86$ GeV/c). As for the cross section for J production, it is known to increase by about three orders of magnitude from the AGS through the ISR range of energies. Do these p_T features and the s-dependence suggest that perhaps the J particle does not have properties expected of a hadron? I do not believe so. In fact, because of the known correlation between mass and p_T in hadronic processes, the value for $<p_T>$ is akin to that expected for a massive hadron.[1] Also, when account is taken of the different masses and Q-values involved, the growth of the J cross section with energy does not differ markedly from the rise observed for K^- or $\bar{p}$ production in pp collisions.[1] (The weaker p_T-dependence of the J/ψ might produce an additional sensitivity to s.) And, finally, recent analyses of photoproduction measurements at SLAC indicate that the extracted J/ψ-nucleon total cross section is ~ 3 mb, which speaks for a hadronic interpretation for the J/ψ particle.

Now I would like to turn to the question of whether we should all feel depressed by the fact that nothing has been found in all the searches for charm-like objects in hadronic collisions. What cross sections should we expect for charm-particle production? Very roughly we can estimate as follows. If charmed particles are produced at rates comparable to production cross sections for other

---

[*]Research supported by the United States Energy Research and Development Administration.

particles having similar mass, then one might expect charm-particle production to be of the same order as antideuteron production ($\sim 1\,\mu b$ at Fermilab energies). Another naive estimate can be obtained by assuming that the production ratio of J/charm is about the same as ϕ/K. From the observed ratio of ϕ/K production, which is about 1/30, we would therefore expect to observe charm production at the level of $\sim 4\,\mu b$ at Fermilab. The upper limits presented for the production cross sections for charmed particles in two-body decay channels are near the 7 μb level.[2] Searches at the ISR have been far less sensitive. Searches at BNL, although relatively more sensitive than at FNAL, are still at about the level where only optimists would expect to find charm. Consequently, I conclude that better experiments are needed to find charm in hadronic channels. One way to enhance one's chances of finding charm is to capitalize, as some have tried, on the fact that charmed particles must have leptonic or semileptonic decay modes. However, if associated charmed-particle production is at $\lesssim 5\,\mu b$ level at FNAL, then, for $\lesssim 20\%$ semileptonic branching ratios of such objects, at most 5% of the prompt lepton yield in hadronic collisions might be attributable to weak decays of charmed particles. For the future, I suspect that small cross sections and difficult experiments will not deter the feistier amongst us from fighting on, after bleeding but a while, until charm is found in hadronic collisions.

REFERENCES

1. See my previous compilations which appear in the review of S. C. C. Ting at the SLAC Conference on Photons and Leptons (1975).
2. I have assumed that charmed particles decay about 5% of the time into two specific bodies. This is what might be expected for hadronic objects having masses of ~ 2 GeV, and is not inconsistent with the upper limits on charmed two-body decays found at SPEAR. I should also stress that the total cross sections I have been discussing are based on great extrapolations of measurements made in restricted regions of p_T and x; these extrapolated figures could easily be off by factors of two or more.

PROMPT POSITRONS AT AGS ENERGIES- REPORT ON THE PENN-STONY BROOK EXPERIMENT

E.W.Beier, R.Patton, K.Raychaudhuri, H.Takeda,
R.Thern, R.VanBerg and H.Weisberg
University of Pennsylvania, Philadelphia, Pa. 19104[*]

M.L.Good, P.D.Grannis, K.Johnson[†] and J.Kirz
State University of New York, Stony Brook, New York 11794[‡]

ABSTRACT

Prompt positrons are observed at beam momenta of 10, 15 and 24 GeV/c near 90° in the c.m. We find the signal to be largest at the lowest $p_\perp$ measured (0.55 GeV/c), where $e^+/\pi^- > 1 \times 10^{-4}$. For 0.8 GeV/c $< p_\perp <$ 1.5 GeV/c vector meson decays may be sufficient to account for the positrons observed.

The aim of our experiment has been to extend the original observations of prompt lepton production [1-4] to lower energies and transverse momenta. Experimental data available at AGS energies on vector meson production allow us to determine the contribution from this well known source of leptons[5]. We were also hoping to find a threshold within the energy range available to us at Brookhaven. This would have helped in the search for an understanding of the effect. It also would have served as an internal test of the method, by demonstrating that we can see both the presence and the absence of a prompt lepton signal. As we shall see, we found no threshold between 10 GeV/c and 24 GeV/c; therefore it must lie outside this range. We believe that we see a signal, and that the threshold lies below 10 GeV/c.

We chose to look for positrons. To measure the prompt signal, we must
1) Identify the positrons originating in the target.
2) Identify or calculate the contributions to the observed positron sample from trivial, well known sources: K_{e3} decay, Dalitz pairs from $\pi^°$, η, ω decay, and e^+e^- pairs produced by γ-rays in and near the target.
3) Calculate the contribution from the "uninteresting" sources: $\rho, \omega, \phi, J/\psi \to e^+e^-$.

The apparatus is shown in Fig. 1. The beam (1-2 x 10^8 particles/pulse) incident on the liquid hydrogen target consisted of ≈100% protons at 24 GeV/c; 85% protons and 15% pions at 15 GeV/c; and 50% protons and 50% pions at 10 GeV/c. Particles produced near 90° in the c.m. entered the spectrometer, built around a magnet and 4 proportional wire chamber modules. To identify positrons we had 3 threshold Cerenkov counters (signals were required in 2 of the 3), and a lead-glass array. This system gave a hadron rejection measured to be 10^6 leaving a residual contamination of between 3 and 7%. To determine

---
[*] Work supported by the Energy Research & Development Administration
[†] National Fellowship Fund Fellow
[‡] Work supported in part by the National Science Foundation

the positron-hadron ratio most directly, we recorded a sample of hadrons along with the positrons.

The contribution to the e^+ triggers due to K decay can be calculated from the known K production spectrum using the Monte Carlo method. We find that this contributes less than 2×10^{-5} to e^+/π^-, and we subtract it. Hyperon decays contribute much less.

Positrons from Dalitz decays and from pair production by γ-rays pose the most difficult problem. Their contribution to e^+/π^- scales roughly as $p_\perp^{-2}$, and at the small $p_\perp$ of our experiment the extrapolation technique used at higher energy[3] becomes very difficult. The pair contribution to our data with $p_\perp > 0.6$ GeV/c combined is illustrated on Fig. 2. The Dalitz contribution alone amounts to 6×10^{-4}, with the γ-ray conversions adding an amount proportional to the thickness of the radiator, but no less than 10×10^{-4} in our apparatus. The common characteristic of these positrons is that they are members of a pair. By building a pair trap, we can identify almost all of them. The trap is shown on Fig. 3. In most pairs the e^+ and e^- are near each other, and will both traverse the dE/dx counters. We observe a clear separation in pulse height between single tracks and pairs. In particular the fraction of pairs faking a single particle is negligible (<<1%).

Some pairs have larger $e^+ - e^-$ separation due to finite opening angles or multiple scattering. These may miss the $\frac{dE}{dx}$ counters, but will be caught in the guard counter array, which form a frame around the first proportional chambers. The combination of these detectors identify 93 - 98% of the Dalitz pairs and about 99% of the $\gamma \to e^+e^-$ pairs in our usual running conditions. The remaining pairs have very large e^+e^- separations, and must be corrected for using a Monte Carlo calculation[6]. This remaining correction is about 20-40% of the observed raw positron signal at our usual running conditions. With radiator added near the target, however, the relative size of these corrections grows considerably[7]. Nevertheless if the calculations are right we should end up with the same residual signal after subtraction independently of radiator thickness. At the present state of our Monte Carlo program there is a tendency for the residual signal to increase with added radiator. To reflect our uncertainty in the calculations, and the disagreement between radiator data and Monte Carlo we have multiplied the Dalitz pair background by $1 \pm 1/3$ and the external pair background by 2 ± 1.

The ratio e^+/π^- after subtraction of hadron background, K decays, Dalitz and $\gamma \to e^+e^-$ pairs is presented as a function of $p_\perp$ on Fig. 4. Statistical errors are shown by the solid bars. The dashed bars include the errors assigned to the Monte Carlo calculations discussed above. The expected contributions from vector meson decay are also shown, together with the range of uncertainty allowed by the data available[5]. We find that while in the larger $p_\perp$ range ($p_\perp \gtrsim 0.8$ GeV/c), vector meson decay is sufficient to account for the positrons we observe, at lower values of $p_\perp$ there is a significant excess. We have not found any reasonable way to account for this excess from known sources[7].

FOOTNOTES AND REFERENCES

1. J. P. Boymond, et al. Phys. Rev. Letters <u>33</u>, 112 (1974)
2. V. V. Abramov, et al. reported in the Proceedings of the Seventeenth International Conf. on High Energy Phys., London, 1974 V-53.
3. J. A. Appel et al. Phys. Rev. Letters <u>33</u>, 722 (1974).
4. F. W. Büsser et al. Phys. Letters <u>53B</u>, 212 (1974).
5. K. Raychaudhuri and H. Weisberg Phys. Rev. <u>D13</u>, 153 (1976). V. Blobel et al. Phys. Letters <u>48B</u>, 73 (1974), Nucl. Phys. <u>B69</u> 454 (1974), H. Gordon et al. Phys. Rev. Letters <u>34</u>, 384 (1975), D. Crennell et al. Phys. Rev. Letters <u>28</u>, 643 (1972).
6. For this calculation we assume that π°, η, and ω inclusive cross sections can be parametrized in the form $E\, d^3\sigma/dp^3 =$ $A \exp(by^2)\exp(cm_\perp)$, where $m_\perp = \sqrt{p_\perp^2 + m^2}$. The parameters A, b, and c are determined from experimental data, where available (see ref. 5). We assumed that the shape of the η production distribution is similar to that of the K^+, with a total inclusive cross section given by the upper limit of K. Jaeger et al. Phys. Rev. <u>D11</u>, 1756 (1975).
7. For thin radiator the number of $\gamma \to e^+ e^-$ pairs is proportional to the radiator thickness, X; while multiple scattering which is responsible for the bulk of the losses scales as $X^{1/2}$. The correction is a combined effect of these, therefore it is not linear in X, and does not lend itself to a simple extrapolation procedure.
8. We note that the signal has a $p_\perp$ dependence not unlike that of the background from Dalitz and $\gamma \to e^+ e^-$ pairs. Our efforts to find a suitable mistake in the background calculation, that would allow us to increase the background and explain the signal have not succeeded.

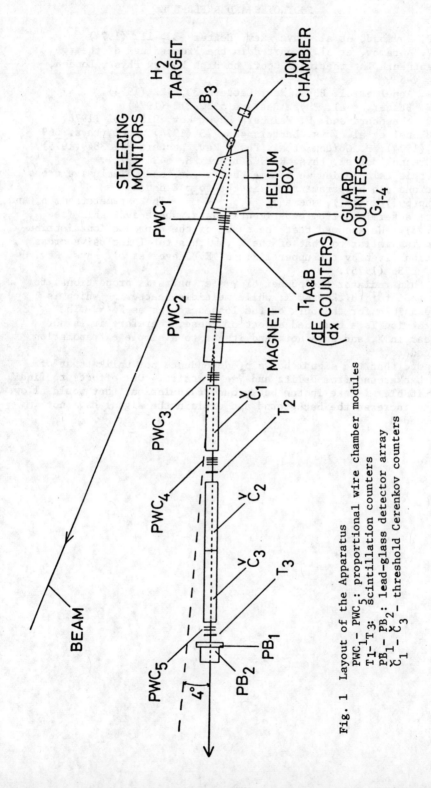

Fig. 1 Layout of the Apparatus
PWC$_1$ - PWC$_5$: proportional wire chamber modules
T$_1$-T$_3$: scintillation counters
PB$_1$- PB$_2$: lead-glass detector array
Č$_1$- Č$_3$: threshold Čerenkov counters

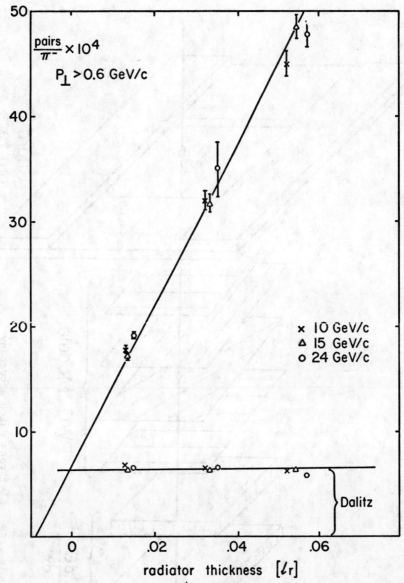

Fig. 2: Points with error bars: e^+e^- pairs/π^- as a function of the thickness of the material in front of the first PWC module. This ratio is obtained from pairs and hadrons recorded along with prompt positron candidates, after detection efficiency and radiative corrections have been applied. All data with $p_\perp > 0.6$ GeV/c are combined. Points without error bars indicate that fraction of the total ascribed to Dalitz pairs based on our detection efficiency calculation at each thickness.

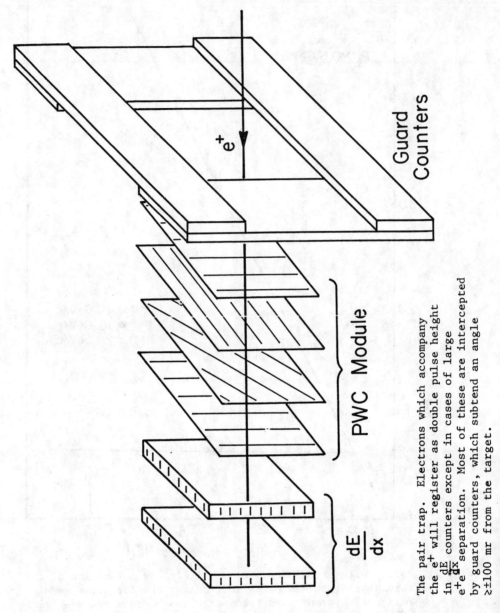

Fig. 3 The pair trap. Electrons which accompany the e^+ will register as double pulse height in $\frac{dE}{dx}$ counters except in cases of large e^+e^- separation. Most of these are intercepted by guard counters, which subtend an angle $\gtrsim \pm 100$ mr from the target.

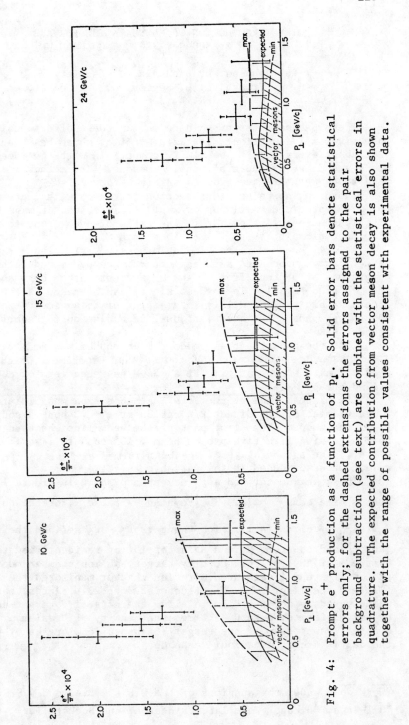

Fig. 4: Prompt e^+ production as a function of $p_\perp$. Solid error bars denote statistical errors only; for the dashed extensions the errors assigned to the pair background subtraction (see text) are combined with the statistical errors in quadrature. The expected contribution from vector meson decay is also shown together with the range of possible values consistent with experimental data.

DIRECT PRODUCTION OF MUONS IN THE FORWARD DIRECTION BY 300-GeV PROTONS ON URANIUM[†]

Rol Johnson, D. Buchholz, O. Fackler,
H. J. Frisch, S. L. Segler, M. J. Shochet,
and R. L. Sumner[*]

The experiment[1] described here was a modest effort designed to answer a simple question: Is the μ/π ratio, which has been measured just about everywhere[2] to be $\sim 10^{-4}$, compatible with the idea that the muons are decay products of massive mesons? The method and direction used to answer this question is straight-forward.

The pion inclusive cross-section increases with decreasing $p_\perp$. For a dileptonic decay ($x \to \mu\mu$ or $x \to \mu\nu$) of an object of mass M, the $p_\perp$ distribution peaks at $p_\perp \simeq M/2$. The ratio of such muons to the pion flux must diminish at $p_\perp$ much less than M/2.

As will be shown, the μ/π ratio does not diminish for $p_\perp \gtrsim 400$ MeV/c. Known processes such as vector meson decay and Bethe-Heitler coversions of photons from π^0 and μ^0 are estimated to be about 20% of the measured signal. The excess signal can be taken as a hint of something exciting or mildly interesting or as an indication that something mundane has been left out, depending on the sobriety of the reader.

That the experiment was modest is a euphemism; most of the equipment was usurped. Of course, any experiment which uses a finite amount of time at a ~$40M/year accelerator cannot be called cheap.

The M1 beam line at the Fermilab Meson Area was violated to allow a measurement of the direct μ/π ratios at x = .3 and .5 and very small $p_\perp$ of the muons. The first part of the beam line was used to transport a quasi-diffractive proton beam ($\sim 10^8$/ppp) to a variable density target. The second stage of the beam formed a ~1μsr spectrometer to select muons produced in the variable target.

The spectrometer had a $p_\perp$ acceptance shown in Figure 1. The momentum bite was $\frac{\Delta p}{p} \sim \pm 7\%$. The Feynman variable x for the data here is $\sim \frac{P_\mu}{P_p} = \frac{P_\mu}{300}$. Six scintillation counters $B_1 - B_6$ defined the muon trajectory and counters μ_1 and μ_2 after 15' of Fe identified the muon. Upstream of the variable density target, an ion chamber monitored the proton flux and a segmented wire ion chamber monitored the beam profile and steering. The target consisted of 23 U plates, each 1" thick, which moved as an accordion to allow the effective target density to be varied from 0.88 to 0.25 times the density of Uranium. Ten feet of steel downstream of the target (6 absorption lengths of U) absorbed the debris from the hadronic cascade.

[†] Work done under the auspices of the U.S.E.R.D.A. and N.S.F.
[*] Cal-Tech, Chicago, Fermilab, Princeton, and Rockefeller.

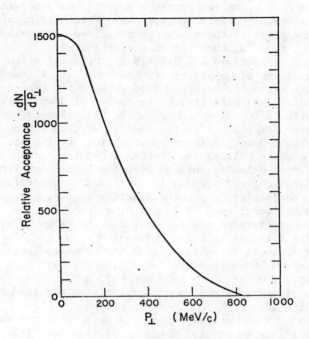

Figure 1

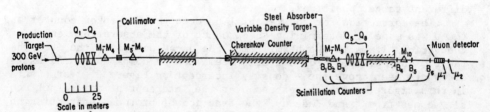

Figure 2

Upstream of the Uranium target a 200' vacuum pipe was modified to be a 0.1 atmosphere He-filled threshold Cherenkov counter to identify muons and pions in the proton beam incident on the variable target. Figure 2 shows a schematic of the apparatus.

The event trigger was $B \equiv B_3 B_4 B_5 B_6$. The pulse heights of the two counters downstream of and nearest to the U target, B_1 and B_2, as well as the Cherenkov counter were recorded on magnetic tape. The data analysis consisted of determining the number of downstream muons not associated with an incoming π or μ in the Cherenkov counter (C), per incident proton for each effective density of the Uranium target. A small, but troublesome, efficiency for protons at the one photoelectron level required raising the Cherenkov counter threshold in the off-line analysis which in turn reduced the efficiency from 95% to 81%. The photoelectron distribution was used to determine the 19% inefficiency correction; since the fraction of triggers caused by beam muons was in the worst case ~0.25 and usually ~0.10, the Cherenkov counter was entirely adequate to determine the rates of proton-induced muons from the U target to a few percent.

These rates were normalized to an ion chamber upstream of the U target. Figure 3 shows the data for three different settings of the downstream spectrometer, $P_\mu = \pm 90$, and $+150$ GeV/c. These correspond to Feynman x values of $x = .3$ and $.5$ for either positive or negative muons as indicated.

The open circles correspond to the data after corrections for Cherenkov inefficiency and for Cherenkov accidentals (<5%). The closed circles with error bars are the same data after corrections for acceptance (5% at the target open position) and for material in the beam between the Cherenkov counter and the Uranium target. The acceptance correction was determined using a Monte Carlo calculation which included the multiple scattering of the muons in the U and Fe. This correction was checked by using a 6" U target in place of the variable density target and absorber to determine the hadron flux for different target positions.

The correction for the material between the Cherenkov counter and the U was made by determining the product of the absorption length of the various items in the beam, times the decay path from the item to the U. This product, for all such items can be compared to the excess of events seen above the ∞ density intercept on Figure 3. For events in this figure, where the excess over the intercept at any target density comes from π and K decay, the fraction of protons interacting in the 6 absorption lengths of U is ~100%. The average π and K decay path is the effective absorption length of the target. The size of the upstream material correction also depends on whether the target is open or closed because the effective decay paths for the upstream items change. The size of the corrections can be seen from the figure. As a check on the sensitivity of the final μ/π ratios on this correction, one can calculate the change in the answer for different assumed values of the absorption length of the U target. A 20% change in the absorption length changes the μ/π ratio by less than 5% in all cases. Since the masses, positions and absorption lengths of all items are known to better than 10%, this systematic error is negligible in comparison to the statistical errors.

The infinite density intercepts of the lines on Figure 3 correspond to the prompt muon production. At infinite density the π and K's have no time to decay and the source of the muons must be some other mechanism. The slopes of the lines are then due to increasing contributions from π and K decay as the target density is decreased. In fact, given the π and K inclusive spectra as a function of x, it is possible to determine the pion and kaon flux per incident proton from the slopes on Figure 3. The intercept/slope ratio is then proportional to the μ/(π+K) ratio. The proportionality constant is determined by integrating the appropriate inclusive spectra. To determine the inclusive spectrum, the 6" U target was used with the variable target and absorber lowered out of the beam. In the interval $.3 < x < .7$ the negative spectrum fit $\frac{d\sigma}{dx} \cong (1-x)^4$. Since the positive spectrum is dominated by protons, the direct measurement would have involved a large correction. So the $\frac{\pi^+}{\pi^-}$ ratios of Aubert et al.[3] were used to determine a π^+ spectrum from the measured negative spectrum. The π^+ spectrum used was $\frac{d\sigma}{dx} \cong x(1-x)^4$. For K^+ and K^-, the spectra used for the calculation were steeper by a factor of x and less steep by a factor of x than the respective pion spectra of the same sign. The spectra and relative K/π ratios were taken from the data of Aubert et al.

Table I shows the final results for the 3 direct μ/π ratios. The calculated proportionality constants are implicit in the figure. In general, a change in the assumed π spectra of a factor of x changes the μ/π ratio by about 20%.

Contributions to the direct lepton signal from vector meson decay as well as Bethe-Heitler conversions from photon decay of π^0 have been calculated. At small $p_\perp$, massive $V^0 \rightarrow \mu\mu$ decays do not contribute significantly to the direct μ rate. For the acceptance of the apparatus and inclusive ρ^0, ω^0, and ϕ^0 data measured at 24 and 205 GeV[4], the Monte Carlo calculations indicate that

$$\left.\frac{\mu^-}{\mu^-}\right)_\rho \cong \left.\frac{\mu^-}{\pi^-}\right)_\omega \cong 1.0 \times 10^{-5},$$

and the ϕ^0 contribution is negligible.

For various parameterizations of the π^0 spectra, the $\frac{\mu^-}{\pi^-}$ ratio from the Bethe-Heitler process is about 1.0×10^{-5}. More significant is the fact that there are some relevant data on this subject. Calibrations of Pb-glass Cherenkov counters at Fermilab with similar s and x show that the $\frac{e^-}{\pi^-}$ rate is about 0.3 for 100% of the photons converting. Scaling by the square of the ratio of lepton masses implies a $\frac{\mu^-}{\pi^-}$ ratio due to the Bethe-Heitler process of 0.75×10^{-5}, in good agreement with the Monte Carlo calculation.

The measurement of the $\frac{\mu^-}{\pi^-}$ ratio at x = .3 for p $\gtrsim$ 300 MeV/c is

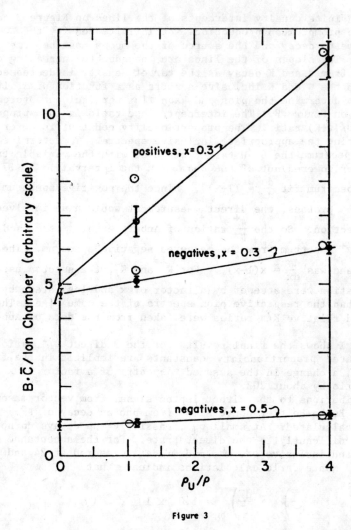

Figure 3

Table I - The corrected intercept to slope ratios and the calculated μ/π ratios at $x = .3$ and $.5$ for negative muons and $x = .3$ for positive muons.

sign	P_μ(GeV/c)	intercept (arbitrary units)	slope due to $\pi + K$	slope due to π only (slope$_\pi$)	$\dfrac{\text{intercept}}{\text{slope}_\pi}$	$\mu/\pi \times 10^4$
+	90	4.95 ± 0.51	1.66 ± 0.22	1.28 ± .22	3.88 ± 0.79	0.47 ± 0.10
−	90	4.72 ± 0.18	0.333 ± 0.077	0.303 ± 0.071	15.6 ± 3.7	1.56 ± 0.40
−	150	0.87 ± 0.12	0.085 ± 0.046	0.085 ± 0.046	10.2 ± 5.6	0.38 ± 0.21

about 5 times larger than conventional mechanisms would indicate. The first possibility is that the measurement is wrong. The only other experiment with similar x, s, and p does, indeed, disagree somewhat with our findings. The results of Leipuner et al.[5] are shown in Figure 4 along with the results of this experiment. As can be seen, our μ^- data are larger by a factor of 5, and μ^+ by a factor of 2.5 than those of Leipuner et al. There are some differences in the two experiments which, in principle, could account for the discrepancies.

Table II   -   Differences.

	Leipuner et al.	This Experiment
A dependence	Cu target	U target
E_{inc}	400 GeV	300 GeV
x acceptance	minimum range requirement	$\frac{\Delta p}{p} \sim \pm 7\%$
$p_\perp$ acceptance	broader, due to more multiple scattering	Figure 1

To the extent that our Bethe-Heitler and vector meson calculations can be applied to the experiment of Leipuner et al., one might judge that their results near x = .3 are explainable. Ours seem not to be.

It is amusing to note that the fraction of the "explained" contribution relative to the total μ^-/π^- ratio estimated by Leipuner et al. (~20%) is the same as for the experiment reported here. Since the data of the two experiments differ by a factor of 5, this implies that the calculations are different by a factor of 5.

What is needed is an experiment which would simultaneously measure the direct μ/π ratios and the contributions due to pairs from vector meson decays and Bethe-Heitler conversions.

REFERENCES

[1] D. Buchholz et al., preprint FERMILAB-Pub-75/88-EXP (Dec. '75) and to be published in Phys. Rev. Lett.

[2] J. P. Boymond et al., Phys. Rev. Lett. $\underline{33}$, 112 (1974).
F. W. Busser et al., Phys. Lett. $\underline{53B}$, 212 (1974).
J. A. Appel et al., Phys. Rev. Lett. $\underline{33}$, 722 (1974). etc.

[3] B. Aubert et al., preprint FERMILAB-Conf-75/31-EXP 7300.001.

[4] V. Blobel et al., Phys. Lett. $\underline{48B}$, 73 (1974).
P. Singer et al., preprint ANL-HEP-PR-75-48.

[5] L. B. Leipuner et al., Phys. Rev. Lett. $\underline{35}$, 1613 (1975).

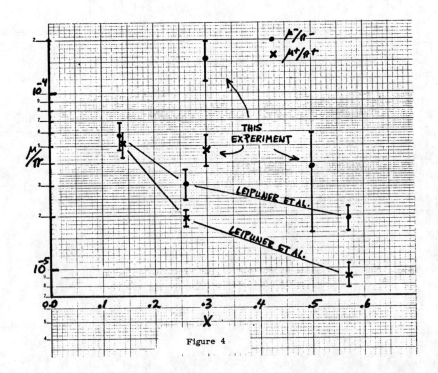

Figure 4

REVIEW OF CERN - COLUMBIA - ROCKEFELLER - SACLAY EXPERIMENT

Samuel L. Segler
The Rockefeller University, New York, N. Y. 10021

ABSTRACT

This paper reviews the results on direct electron production in p-p collisions at the CERN ISR from the CCRS collaboration. In particular, the cross sections are presented at $\sqrt{s}$ = 22.5, 30.6, 44.8, 52.5 and 62 GeV and compared with the π^0 cross section. Correlations with hadrons are given and compared with those for pions. Finally, a comparison is made with the contribution to single e's from the J/ψ and with upper limits obtained for low mass e^+e^- pairs, i.e., $.3 \leq M_{ee} \leq .8$ GeV/c^2.

The CCRS experiment studied the two reactions

$$p + p \to e + \text{anything} \tag{1}$$

$$p + p \to e^+ + e^- + \text{anything} \tag{2}$$

I will review the results relevant to process (1). Table I gives the luminosity for each value of $\sqrt{s}$. These results are an update of the results presented at the SLAC conference last August[1].

TABLE I - Total luminosity at each $\sqrt{s}$ value

$\sqrt{s}$ (Gev)	30.6	44.8	52.7	62.4
Integrated Luminosity	2.86 x 10^{35}	4.83 x 10^{35}	8.95 x 10^{35}	1.25 x 10^{35}

The apparatus has been described in detail elsewhere[2,3]. It consisted of two magnetic-lead glass spectrometers located at ± 90° with respect to the incident ISR beam directions. The essential features were pulse height measurement close to the beam pipe, a magnet for momentum measurement, lead glass for measuring the electron energy, and a Cerenkov counter to reject hadrons. Arm I had a lead-scintillator arrangement rather than lead glass.

It is useful to review here the backgrounds which contributed to the events of interest. These were: i) γ-ray conversion in the beam pipe; ii) Dalitz decays from π^0 and η^0 mesons; and iii) hadrons which interacted in the lead-glass as well as giving a large pulse height in the Cerenkov counter. Requiring one particle in the hodoscope counter next to the beam pipe suppressed backgrounds (i) and (ii), while requiring the momentum and energy to agree as well as the correct Cerenkov counter cell to fire suppressed background (iii).

The γ-ray conversions were measured by inserting additional converter and then extrapolating the yield as a function of converter thickness to zero converter[2]. The hadron background was measured by making the same requirement as for electrons except that the wrong Cerenkov cell be set. One could then compare the distributions of events as a function of momentum divided by lead-glass energy for accepted events, pure electrons, and hadrons as shown in Figure 1. Normalizing curves (b) and (c) to add up to (a) gives the hadron contribution.

The contribution from Dalitz decay must be calculated by Monte Carlo and we have recently redone this including the complete contribution from π^0 and η^0, assuming η^0/π^0 is 0.55 as a function of p_T^*. The Dalitz pairs have greater acceptance than do the conversion pairs because of their wider distribution in opening angle, allowing one member of the pair to miss the first hodoscope counter. The mass distributions for both π^0 and η^0 Dalitz pairs are shown in Figure 2. The distribution used for this calculation was[4]:

$$\frac{d^2\Gamma_{\gamma ee}}{dx d(\cos\theta_m)} = \Gamma_{\gamma\gamma} \frac{\alpha}{2\pi} \frac{1}{x} (1 - \frac{x^2}{m^2})^3 \sqrt{1 - \frac{4m^2}{x^2}} (1 + \frac{4m^2}{x^2}) *$$

$$* [1 + \frac{x^2 - 4m^2}{x^2 + 4m^2} \cos^2\theta_m] \quad (3)$$

where
- x = pair mass
- m = mass of electron
- M = π^0 or η^0 mass
- θ_m = c.m. angle of positron with respect to virtual photon direction

The contribution, then, to the accepted events is given in Figure 3, expressed in terms of $\frac{e^+ + e^-}{2\pi^0}$ vs. p_T^*. The steep rise below ~1.0 GeV/c makes it difficult to extract the direct electron signal cleanly at low p_T (≤ 1.0 GeV/c).

TABLE II - Background to single electrons

Background	% of accepted events
Charged Hadrons	18.6 ± 1.9
γ - conversions	8.3 ± 2.4
Dalitz Decays	≃ 20
Total	~ 46.9

Table II gives the percentage contributions for all three backgrounds to the accepted events for $p_T > 1.3$ GeV/c. The preliminary data presented at SLAC had subtracted from it a Dalitz contribution of ~ 6%. Hence the overall cross section above $p_T = 1.3$ GeV/c is now lower by about 30%. As there is no asymmetry within errors between e^+ and e^-, we average the two to present the cross sections.

The electron cross sections are shown in Figure 4. The dotted lines indicate the corrected cross sections when the total contribution from Dalitz decay is subtracted. Integrating the cross sections from $p_T \geq 1.3$ GeV/c and plotting the result yields Figure 5. As can be seen, the cross section rises over the ISR energy range by a factor of 2.5 to 3.

We have also compared the behavior of the electron cross section with that of pions as a function of $\sqrt{s}$ and p_T. The most direct method is to take the ratio of "direct" electrons to selected conversion electrons, obtained by requiring two particles in the first hodoscope counter. One can then by means of a Monte Carlo convert this ratio to an e/π^0 ratio. To do this, we used the B-S fit for the $\pi^\pm$ cross section over our p_T range and assume $\eta^0/\pi^0 = .55$. The result is given in Figure 6 for the different $\sqrt{s}$ values, plotted as a function of p_T^*. Once again, the dotted lines indicate the value after subtracting the complete Dalitz contribution.

Integrating the cross sections for $p_T^* \geq 1.3$ GeV/c and plotting as a function of $\sqrt{s}$ yields Figure 7. From the plot there is an indication of some $\sqrt{s}$ dependence of the ratio $\frac{e^+ + e^-}{2\pi^0}$.

It is useful to consider also the correlations with hadrons of the accepted direct electron events with those for hadrons. Figure 8 shows the same side result where the number of charged hadrons per interaction is plotted as a function of the p_T of the hadron. An interaction is defined as either a "direct" electron or a selected conversion electron. No difference between the two distributions is seen. The same is shown in Figure 9 for charged hadrons on the opposite side where once again no difference is seen. We can also separate out the different hadrons by time-of-flight and as shown, no difference in the hadron composition is seen within errors.

To attack the question of the source of the electrons, we have calculated the contribution of the ψ/J to the direct e signal. The calculation assumed

$$E \frac{d^3\sigma}{dp^3} = e^{-bp_T}$$

where this form is suggested by the data obtained by us for electron pairs from the ψ/J resonance. Figure 10 shows the result for different values of b. Our conclusion is that in order to explain the signal at 52.7 GeV/c, a value of $<p_T>$ must be chosen which then disagrees with the direct e slope. Hence the ψ/J cannot explain the signal and certainly fails totally below 2.0 GeV/c p_T.

A search was made for same side e^+e^- pairs by requiring two tracks which both satisfy the single particle requirement in the front hodoscope and both set the Cerenkov counter cell bit. Only two events were seen for $M_{ee} \geq 300$ MeV/c². In order to calculate the acceptance and hence upper limits for a discrete mass particle, it was assumed that the particle is produced with a p_T dependence given by the B-S fit[5] and then decays isotropically to e^+e^-. Table III gives the 95% confidence level upper limits where four events should have been seen in a 100 MeV/c² wide bin centered on the particle mass. The resulting fraction of the single e cross section at 52.7 GeV which could be explained is also given. As can be seen, we rule out masses between .3 and about .8 GeV/c². Above that, the acceptance falls off because both leptons are not seen in the apparatus.

TABLE III - Upper limits on same side pairs

Pair Mass GeV/c²	95% Conf. Level $B \frac{d\sigma}{dy}(p_T^* > 1.3)$ (cm²)	Fraction of singles corresponding to this limit
.4	5.54 x 10^{-33}	.064
.5	8.37	.104
.6	1.64	.228
.7	3.45 x 10^{-32}	.412
.8	1.08 x 10^{-31}	1.613
.9	1.77	2.73
1.00	2.65	4.45

$$\frac{d\sigma}{dy}\bigg|_{\pi^{\pm}}(p_T^* > 1.3) = 2.71 \times 10^{-28} \text{ cm}^2$$

If we consider the possibility of a low mass continuum, the same upper limits apply if we take the cross section now to correspond to the integral of the continuum over the .1 GeV/c² wide bin with the center as given in Table III. It can be seen that one can easily explain the signal by adding the contributions from each mass bin. Hence, we do not rule out the possibility of a low mass continuum which could give rise to the direct e's [6].

REFERENCES

1. F.W. Busser et al., Direct Leptons at the ISR II, Proc. of the International Conf. on Lepton and Photon Interactions, Stanford (1975).
2. F.W. Busser et al., Phys. Letters 53B, 212 (1974).
3. F.W. Busser et al., Phys. Letters 56B, 482 (1975).
4. D.W. Joseph, Nuovo Cimento 16, 997 (1960).
5. B. Alper et al., contribution No. 227 to session A3 of the 17th International Conf. on High-Energy Physics, London (1974).
6. G.R. Farrar and S.C. Frautschi, submitted to Phys. Rev. Letters.

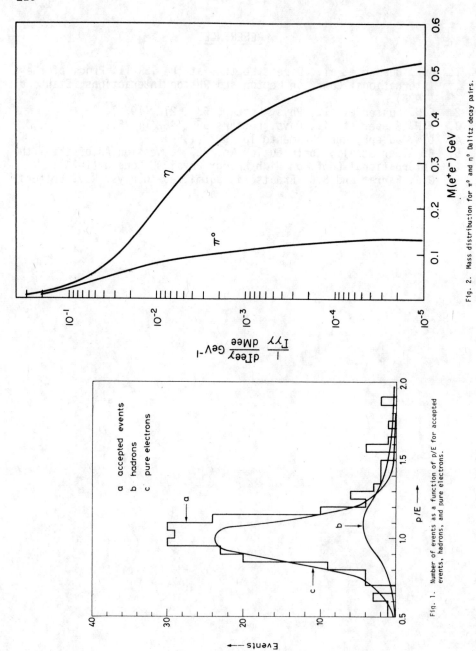

Fig. 2. Mass distribution for π^0 and η^0 Dalitz decay pairs.

Fig. 1. Number of events as a function of p/E for accepted events, hadrons, and pure electrons.

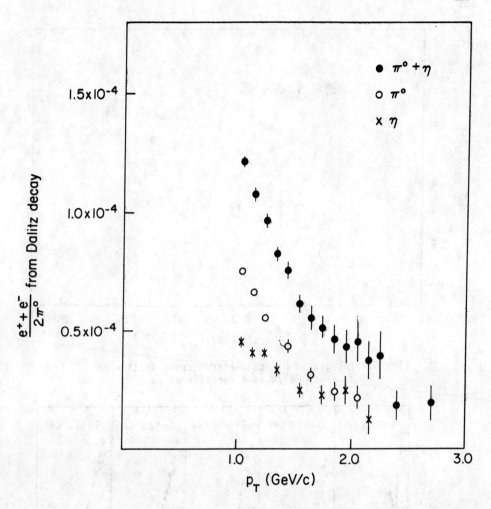

Fig. 3. Contribution of Dalitz pairs to accepted events as a function of p_T.

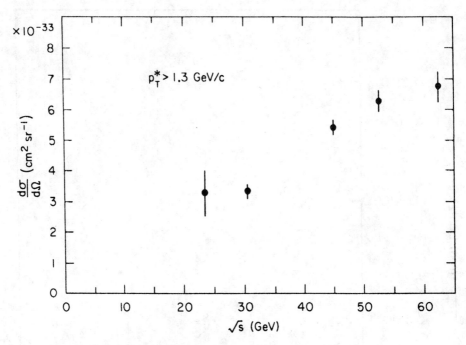

Fig. 5. Integrated direct electron cross section for $p_T > 1.3$ GeV/c as a function of $\sqrt{s}$.

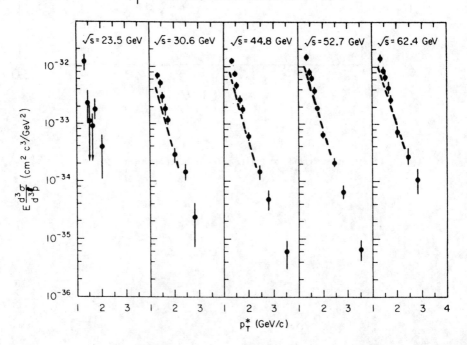

Fig. 4. Charge-averaged invariant cross section for direct electron as a function of p_T, for different values of $\sqrt{s}$.

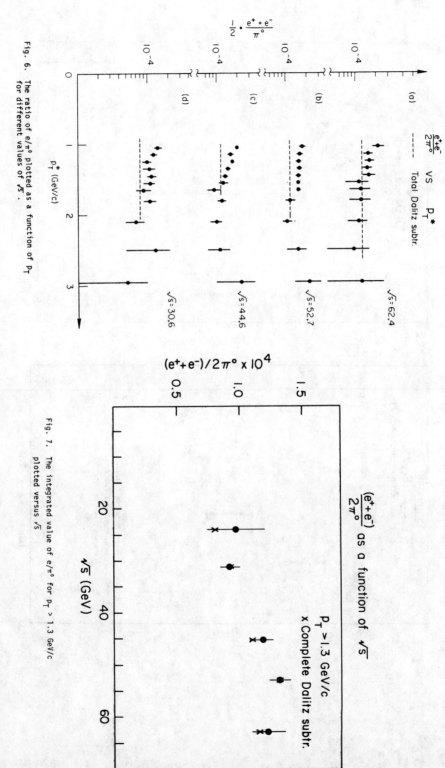

Fig. 6. The ratio of e/π^0 plotted as a function of p_T^* for different values of $\sqrt{s}$.

Fig. 7. The integrated value of e/π^0 for $p_T > 1.3$ GeV/c plotted versus $\sqrt{s}$.

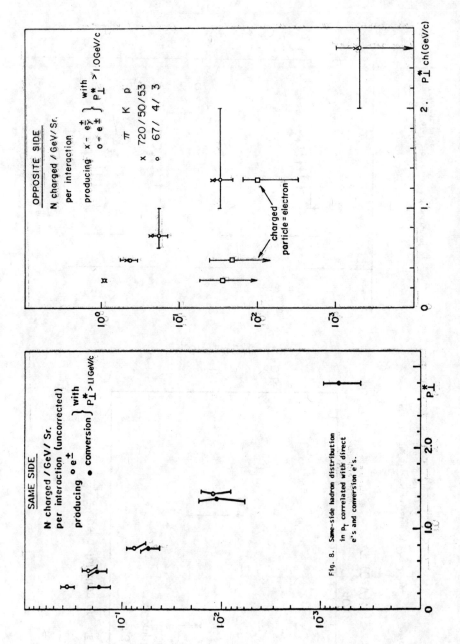

Fig. 8. Same-side hadron distribution in p_T correlated with direct e's and conversion e's.

Fig. 9. Opposite-side hadron distribution in p_T correlated with direct e's and conversion e's.

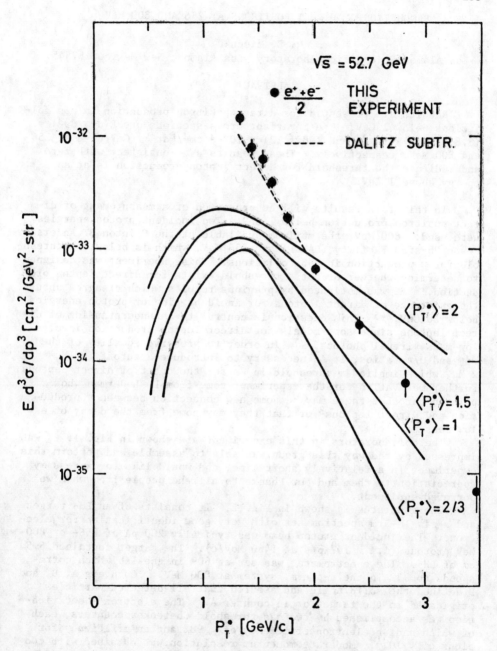

Fig. 10. Charge-averaged contribution of ψ/J to direct electron cross section at $\sqrt{s}$ = 52.7 GeV for different values of $\langle p_T \rangle$ of ψ/J.

PRODUCTION OF DIRECT POSITRONS AT 256 AND 800 MeV[*]

R. E. Mischke

Los Alamos Scientific Laboratory, Los Alamos, New Mexico 87545

ABSTRACT

Results are presented for direct positron production in p-p collisions below 1 GeV. The invariant cross sections are $E(d^3\sigma/dp^3)$ = $(3.8 \pm 5.4) \times 10^{-32}$ and $(-8 \pm 12) \times 10^{-32}$ cm^2/[sr · (GeV/c)^2] at 256 and 800 MeV, respectively. These results are consistent with zero and indicate the threshold for direct lepton production is at an energy above 1 GeV.

In this talk results will be presented of a measurement of direct positron production below 1 GeV. Two incident proton energies were used: 800 MeV, the maximum available at the Clinton B. Anderson Meson Physics Facility (LAMPF) and 256 MeV, which is below the threshold for π production in p-p collisions. This experiment was designed to determine whether or not a threshold exists for direct lepton production.[1] It was motivated by previous results which disagree about the existence of direct leptons for incident pion or proton energies below 30 GeV.[2-5] This question is central to an understanding of direct leptons since some models for direct lepton production involve low-mass virtual photons.[6,7] In order to preserve equality of the e/π and μ/π ratios, it is necessary to introduce a cutoff for M_{γ}^* < 2 M_{μ}, which implies a threshold in $\sqrt{s}$ for the onset of direct leptons. Further motivation for the experiment came from L. Lederman who pointed out that there may be some new connection between π production and direct leptons[8] or that they may come from the decay of a low-mass particle.[9]

The collaborators in this experiment are shown in Fig. 1. I was impressed by the way this group was able to assemble and perform this experiment in a relatively short time, and would like to express my appreciation to them and our thanks to all the people from whom we borrowed equipment.

The apparatus is shown in Fig. 2. It consists of an LH$_2$ target and a single-arm spectrometer with very good identification for electrons. The incident proton beam was typically 200 pA of 256- or 800-MeV protons (i.e., 10^9 p/s at 1460 MeV/c). The target contained 6.5 cm of LH$_2$. The spectrometer was set at 60° in the lab which corresponds to 93° in the p-p c.m. system at 800 MeV. Counters A, B, and D defined the solid angle and assured that all detected particles originated in the target or in counter A. The electron identification was accomplished by two gas-threshold Cherenkov counters, each of which had an electron efficiency of ~ 98% and an efficiency for pions of ~ 10^{-4}. Moderate momentum resolution was obtained with two banks of hodoscope counters. They were constructed from Pilot 425

---

[*] Work performed under the auspices of the U.S. Energy Research and Development Administration.

plastic and were insensitive to protons. An additional check of the electron identification was provided by an array of Pb glass shower counters. Figure 3 shows a plot of the energy in the shower counters divided by the momentum determined by the spectrometer. This quantity peaks at 1.0 for electrons and is small for π's. Even though the term electron is used from habit, we actually detected positrons to avoid the problems associated with electrons produced from Compton scattering.

The time available does not permit a detailed discussion of either the data-taking or the analysis. Thus I will proceed directly to a discussion of the results at 256 MeV. Figure 4 shows the yield for 256-MeV incident protons. The four groups of runs correspond to target full and empty and zero or two layers of mylar absorber added between the target and first scintillator. Each absorber or converter was 0.15 g/cm^2 thick. The runs of each type are internally consistent except for the zero absorber target full; no reason has been found for the discrepancies in this set other than a statistical fluctuation. The yield increases when the target is filled or when absorber is added, as expected; but even with target empty there is a considerable electron flux. This is due primarily to $C(p,\pi^0)$ reactions in the target walls where π's are produced coherently from the carbon, giving rise to a flux of γ's. Some of these γ's convert in the available material and, when the target is full, some additional γ's convert in the H_2. Thus there will be a difference between target full and empty, even if nothing is produced in the hydrogen directly. Taking the difference between target full and empty for the two-absorber data and correcting for conversions in the H_2, we obtain a laboratory differential cross section of $d\sigma/d\Omega dp = (5.6 \pm 4.5) \times 10^{-36}$ cm^2/sr-MeV/c. The momentum range included is from 70 MeV/c (the lower limit of the spectrometer) to ~ 140 MeV/c (the kinematic limit at this energy); the mean momentum is 85 MeV/c. The normalization of the data is obtained from an ionization chamber which monitored the incident proton beam and a Monte Carlo calculation of the acceptance of the spectrometer.

The calculated contribution from p-p bremsstrahlung[10] for this incident energy and range of detected electron energies is $d\sigma/d\Omega dp = (2.4 \pm 0.7) \times 10^{-36}$ cm^2/(sr · MeV/c). The net signal, converted to an invariant cross section is

$$E \frac{d^3\sigma}{dp^3} = (3.8 \pm 5.4) \times 10^{-32} \text{ cm}^2/[\text{sr}(\text{GeV/c})^2] .$$

The 800-MeV data are shown in Fig. 5. The three curves correspond to zero, two, and four added absorbers. The plotted data have the target-empty yield subtracted and include corrections for dE/dx and radiative losses in the front half of the spectrometer. From the data it is clear that we see electrons--an exponentially increasing number of them as the momentum decreases. The problem is to determine whether or not all of these come from π^0 Dalitz decay and pair production. A straight line is fit to the zero, two, and four absorber yield for each momentum bin to determine the electron yield per added absorber. This information is used to subtract the contributions to the electron yield with zero added absorber due to pair production in the target and first scintillator and due to Dalitz decay. Dalitz decay is conceptually just like a fixed converter, but a Monte Carlo program is used to calculate the ratio of Dalitz to pair production contributions at each momentum. These subtractions are illustrated schematically in Fig. 6 and show that the direct lepton component is the excess, if any, after the subtractions have been made.

The results, shown in Fig. 7, are consistent with zero. Combining all the data into a single point with a weighted average momentum of 318 MeV/c ($p_\perp$ = 275 MeV/c), the result is $d\sigma/d\Omega dp$ = $(-25 \pm 38) \times 10^{-36}$ cm^2/(sr · MeV/c). This translates into an invariant cross section of

$$E \frac{d^3\sigma}{dp^3} = (-8 \pm 12) \times 10^{-32} \text{ cm}^2/\text{sr(GeV/c)}^2 .$$

Thus it is clear that we do not observe any evidence for direct leptons at either 256- or 800-MeV incident protons. The next question is how to compare these results to those at much higher energies. We can quote a limit on the ratio e/π, but since we are not in an asymptotic region, it might be thought that pion yields are much different due to resonance production. Actually, if you discount the falloff due to kinematics, the π invariant cross section for π production is remarkably insensitive to $\sqrt{s}$ (Fig. 8). Thus the e/π ratio has a similar interpretation at all energies in the appropriate region of $p_\perp$. Even so, we prefer to quote our results in terms of the invariant cross section. The comparison to existing data[11-13] is shown in Fig. 9. Our upper limits are "orders of magnitude" below the universal curve connecting the high-energy results.

We believe that direct leptons of the intensity observed at high energies are not present below 1 GeV. This implies the existence of a threshold, and the Penn-Stony Brook experiment indicates that this threshold lies between 1 and 10 GeV/c.

REFERENCES

1. The term direct lepton, defined in previous talks, refers to leptons (either electron or muon) initially observed at large transverse momenta in high-energy hadron-hadron collisions in excess of those which arise from π and k decay.
2. Penn-Stony Brook experiment presented at this conference by J. Kirz.
3. R. C. Lamb et al., Phys. Rev. Letters 15, 800 (1965).
4. R. Burns et al., Phys. Rev. Letters 15, 830 (1965).
5. K. Winter, Phys. Letters 57B, 479 (1975).
6. M. Duong-van, SLAC preprint PUB-1604 (1975).
7. G. Farrar and S. Frautschi, Cal. Tech. preprint CALT-68-518 (1975).
8. L. Lederman, Proceedings of the Sixth International Conference on High-Energy Physics and Nuclear Structure, AIP Conference Proceedings No. 26 (1975).
9. L. Lederman and S. White, Phys. Rev. Letters 35, 1543 (1975).
10. G. Bohannon, private communication.
11. CHORM collaboration, presented at this conference by C. Rubbia.

12. CCRS collaboration, presented at this conference by S. Segler.
13. J. P. Boymond et al., Phys. Rev. Letters 33, 112 (1974).

COLLABORATORS

A. BROWMAN, LASL	L. AUERBACH, TEMPLE
J. FRANK	V. HIGHLAND
C. HOFFMAN	K. McFARLANE
H. HOWARD	
R. MISCHKE	
D. MOIR	
D. NAGLE	

Fig. 1. List of collaborators for this experiment.

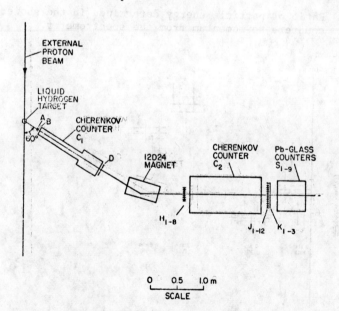

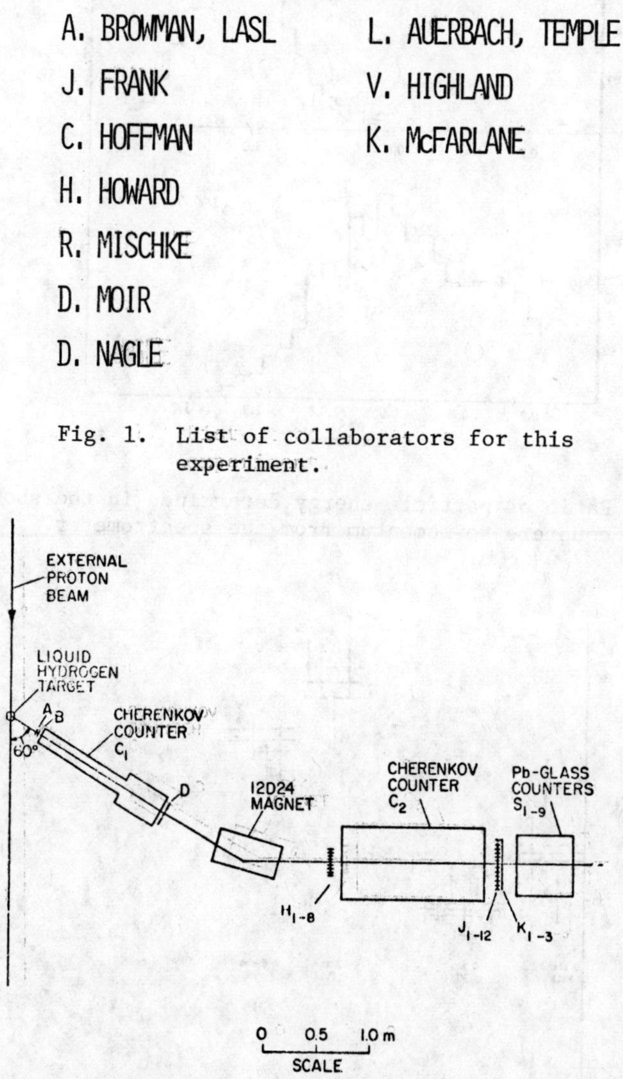

Fig. 2. Schematic diagram of the apparatus. Counters A, B, and D are plastic scintillators, C_1 and C_2 are gas threshold Cherenkov counters, H, J, and K are plastic Cherenkov counters.

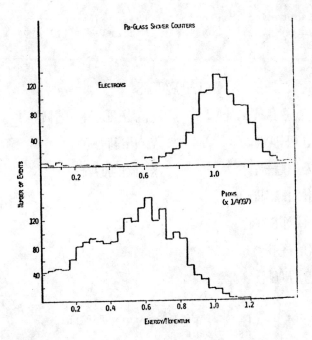

Fig. 3. Ratio of particle energy determined in the shower counters to momentum from the spectrometer.

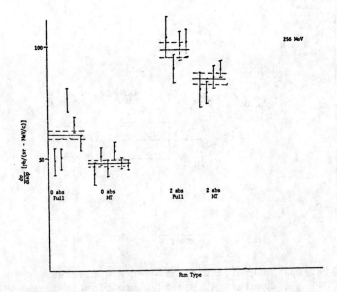

Fig. 4. Electron yield for different conditions for the 256 MeV data.

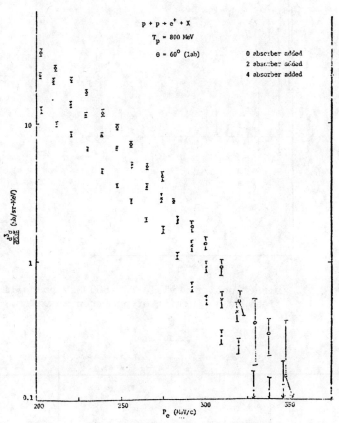

Fig. 5. Electron yield at 800 MeV for 0, 2, and 4 added absorbers.

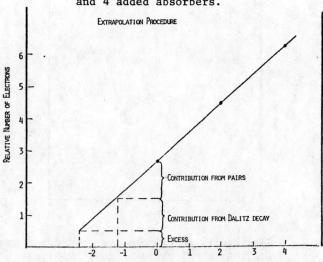

Fig. 6. Schematic illustration of the procedure for extrapolation to zero effective material.

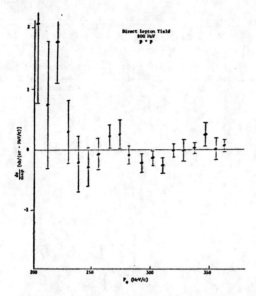

Fig. 7. Direct positron yield for 800 MeV pp collisions as a function of positron momentum.

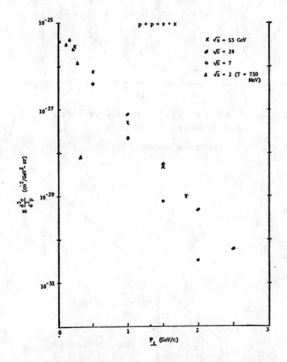

Fig. 8. Sample data on π inclusive production showing that the invariant cross section is insensitive to √s.

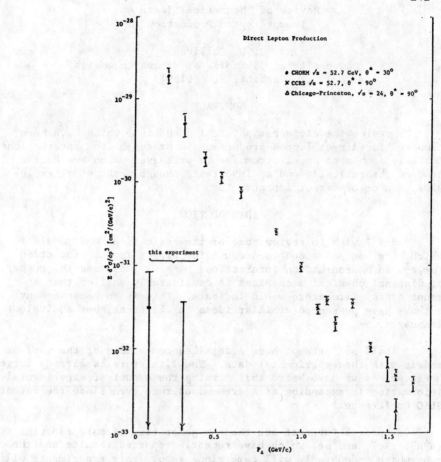

Fig. 9. The results of our experiment plotted together with the invariant cross section for direct leptons from high energy data.

Review of Theoretical Ideas on
Prompt Lepton Production*

J. D. Sullivan
University of Illinois at Urbana-Champaign
Urbana, IL   61801

ABSTRACT

We review the different physical mechanisms which have been proposed for direct lepton production in hadron-hadron interactions. Emphasis is placed on electromagnetic processes which are better studied theoretically and an important, though perhaps not exclusive, source of direct leptons.

INTRODUCTION

Today I wish to review some of the basic ideas and models which have been advanced to account for direct lepton production ($\mu^\pm, e^\pm$) in hadron-hadron interactions. As we shall see the number of distinct physical mechanisms is considerably smaller than a count of the literature would indicate. This is so because many authors have presented similiar ideas in different, but equivalent languages.

I shall not attempt here a detailed comparison of the various models with the experimental data. The literature is already large. Moreover, as we have heard this morning the amount of experimental information is expanding at a tremendous rate even since the recent SLAC Conference.[1]

Of special interest are the new generation of pair experiments[2] ($\mu^+\mu^-$, e^+e^-, and μe) which have recently reported results and the many which undoubtedly will come along soon. Pair experiments will provide definitive information about the source of direct leptons.

One cannot help but recall the analogous situation in strong interaction dynamics. Many different models (e.g. multiperipheral, fireball, etc.) were able to successfully fit one particle spectra. Only by detecting two or more hadrons and thereby obtaining information about correlations was it possible to select those mechanisms which are dominant in nature.

MECHANISMS

Direct lepton production could, perhaps, be coming from totally new and unexpected phenomena. However, given the large uncertainty and gaps in our present knowledge, it seems more productive to adopt the conservative point of view that the lepton couplings

---

*Work supported in part by the National Science Foundation under Grant No. MPS 75-21590.

are the conventional ones. Namely:

(1) an additive conservation law separately for μ type and e type leptons,

(2) standard QED couplings for μ and e, and

(3) universal μ - e coupling for the weak interactions.

Thus, the occurrence of a charged lepton ℓ in the final state of a hadron-hadron interaction must be accompanied by either another charged lepton $\bar{\ell}$ or by the appropriate neutrino $\nu_{\bar{\ell}}$. We can divide the interesting possibilities into four broad categories:

I. Weak decay of resonant states ($W^{\pm}$, Z^0, ?) formed by weak interaction processes. (We include here, also, the electromagnetic analog[3] (B^0) should such heavy photon states exist.)

II. Weak decay of resonant hadronic states (π, K, D, F, D^*,...) formed by strong interaction processes.

III. Electromagnetic decay of resonant hadronic states (π^0, η; ρ^0, ψ/J, ψ',...) formed by strong interaction processes.

IV. Production of a virtual photon continuum by electromagnetic processes.

Other combinations of production and decay schemes are possible but seem hopelessly small (e.g. the weak analog of IV). Also omitted in the list above is the interesting possibility that direct electrons and muons are the decay products of heavy leptons $L^{\pm}$ produced by one or more of the above mechanisms. I am unaware of any attempts to propose this as a dominant mechanism, however.

WEAK MECHANISMS

Mechanism I used to be the sole motivation of the lonely few who engaged in direct lepton experiments. However, the negative results of all experimental searches[4] and the theoretical development[5] of attractive (though not yet proven) unified theories of the weak and electromagnetic interactions suggesting $m_{W^{\pm}}$, m_{Z^0} = $O(100$ GeV/c^2) have caused interest in (I) to be postponed until the next generation of superaccelerators. Mechanism II is viable only for resonant states which are stable under the strong and electromagnetic interactions since otherwise the (semi-) leptonic branching ratio would be exceedingly small. Pion, kaons, and etas are so well-known and such copious sources of leptons that all experimentalists remove these contributions in one way or another and report only the remaining direct (i.e. unknown and therefore interesting) lepton yield.

Interest currently focuses on the possibility of significant contributions from the charmed mesons M_c - either the pseudoscalars, D and/or F, or the vector mesons D^* and/or F^*, predicted by the popular four-quark model.[6] (Similar states and more are predicted by the various six-quark models[7] which have been advanced recently.) The masses of these mesons, as calibrated by charmonium calculations[8] and more directly by the rise in σ ($e^+e^- \to$ hadrons) observed at SLAC-SPEAR[9] are expected to be $M_c \approx 2$ GeV/c^2. Hence, associated production $M_c \bar{M}_c$, $B_c \bar{M}_c$, etc. at Fermilab and CERN-ISR is kinematically possible over a large region of phase space. In the central rapidity region $M_c \bar{M}_c$ is expected to dominate yielding equal ℓ^+ and ℓ^- signals. Only in the fragmentation region where $B_c \bar{M}_c$ might become important (by analogy to ΛK, ΣK) can we expect a departure from charge symmetry in the single lepton spectrum.

It is very difficult however, to make a quantitative estimate of the lepton yield from charmed meson (or baryon) production. This is so because there is considerable uncertainty in estimates of production cross section, (semi-)leptonic branching ratios, and even the masses of the hadronic system X which is occurring in the decay $M_c \to X + \ell + \bar{\nu}_\ell$. Present production estimates are based on either simple scaling laws applied to $K\bar{K}$, pp, and ψ/J data[10] or theoretical estimates[11] based on the idea that high energy - high mass behavior is susceptable to perturbation theory calculations in the colored gauge theory of quarks (asymptotic freedom). Only experiment can tell if these approaches are reliable guides. Typical theoretical estimates at Fermilab and ISR energies range from 0.1 µb - 10 µb for inclusive $M_c \bar{M}_c$ production.

The branching ratio question has received a somewhat more quantitative treatment.[12] The semileptonic branching ratio for the decay $M_c \to X + \ell + \bar{\nu}_\ell$ (for $\ell = \mu$ or e) is estimated to be 1 - 25%. It is reasonable to expect that this branching ratio will be more tightly constrained by the observation and analysis of neutrino induced events with two charged leptons in the final state which we will hear about at this conference in the afternoon.

We can also expect progress from another direction. A large part of the spread in the theoretical estimates of the leptonic branching ratio comes from differing assessments (see Ref. 12) of the amount of enhancement of <u>non-leptonic</u> charmed meson decay. Some of the theoretical reasons behind these estimates are being tested at SLAC-SPEAR.[13] There, according to the current view, pair production of $M_c \bar{M}_c$ is occurring at a sizable fraction of the total annihilation cross section above center-of-mass energy 4.1 GeV. These same non-leptonic branching ratios (especially the two-body ones) play a critical role in many of the charm search experiments we have already heard about at this conference.

While it is generally expected that the charmed pseudoscalar mesons D and F will be less massive than the vectors D^* and F^* it has been suggested[14] that perhaps the opposite might occur. If, in fact, the charmed vectors are the lightest, and hence the only ones stable against strong decays, the effect on direct lepton production could be dramatic for the following reason.

The two-body decay D (or F) $\to \mu + \bar{\nu}_\mu$, $e + \bar{\nu}_e$ is severely suppressed for the same angular momentum reasons that the $\pi_{\mu 2}$ rate dominates π_{e2}. However for D^* (or F^*) $\to \ell + \bar{\nu}_\ell$ this suppression no longer occurs and hence direct lepton production at $p_\perp \gtrsim$ 1 GeV/c would be enhanced. Perhaps the best signature for this would be an unexpected high energy tail on the $\nu_e (\bar{\nu}_e)$ and $\nu_\mu (\bar{\nu}_\mu)$ beams at Fermilab.

Barring this interesting but unlikely possibility, charmed meson production will contribute to direct lepton production via three-body decay: $M_c \to X + \ell + \nu_\ell$. Even for small m_X this is a very inefficient mechanism for producing large $p_\perp$ leptons since the transverse momentum of the parent and energy release must be shared among three bodies. Fig. 1 illustrates this for the case of a meson M_c = 2GeV/c² produced according to

$$\frac{d\sigma}{dy d^2 q_\perp} = 0.6 \, \mu b/(GeV/c)^2 \exp(-2q_\perp/(GeV/c)) \tag{1}$$

decaying isotropically via $M_c \to e + \nu_e$ vs $M_c \to K + e + \nu_e$ with a 10% branching ratio for either case.

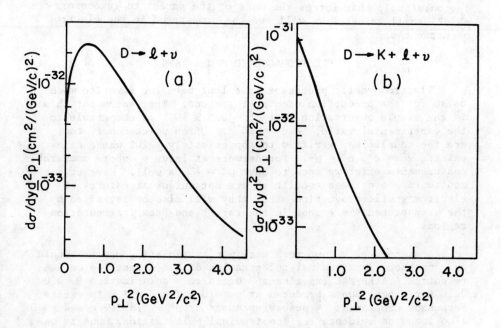

Fig. 1 The transverse momentum spectrum for leptons produced via two-body (a) and three-body (b) of a charmed meson produced according to Eq. (1).

For three body decay we note the distribution peaks at the origin, extends out to $p_\perp \approx 1$ GeV/c and then falls much more steeply at large $p_\perp$ than the distribution for two body decay. It seems very difficult, however, to fit the sharp increase in the ratio e/π observed[16] at CERN-ISR as one goes from $p_\perp \simeq 1.0$ GeV/c down to $p_\perp = 0.2$ GeV/c, the smallest value thus far observed. In particular the ratio of the direct electron yield in Fig. 1 (b) to $\pi^\pm$ data decreases below $p_\perp = 0.5$ GeV/c in contrast to the experimental ratio. Furthermore, any attempt to match the magnitude of the direct electron yield observed in Ref. 16 requires a very large inclusive cross section for charmed particle production ($\sigma(M_c \overline{M}_c) \approx 1$mb if $B(D \to K + e + \nu_e) = 10\%$). Thus we conclude that even though low $p_\perp$ is the most propitious region for seeing direct leptons from charmed particle decay we do not yet have a satisfactory understanding of what is actually happening and must await further work before a verdict can be given.

Finally we note that Lederman and White[17] have observed that all of the direct lepton production data extant at the time of the SLAC Conference was compatable with the decay of a new light mass particle $X^\pm \to \mu^\pm + \nu_\mu$, $e^\pm + \nu_e$ produced at a rate x branching ratio $\simeq 10^{-3}$ that of pion production and $m_X = 100 - 700$ MeV. Putting aside the question of how such a light mass particle could have previously escaped notice, we note that it is basically the kinematics of two-body decay, given the observed $\sqrt{s}$ and $p_\perp$ dependence of the direct lepton signal, that forces the mass of the parent to (uncomfortably?) small values. We will see the same trend in the electromagnetic case.

ELECTROMAGNETIC MECHANISMS

Electromagnetic processes have long been the favorite mechanism for the production of direct leptons. The reason for this is the simple observation that $\alpha^2 \simeq 0.5 \times 10^{-4}$ is comparable to the experimental ratio, $\ell^\pm/\pi^\pm \simeq 10^{-4}$. Often pointed out, too, are the equalities, verified to approximately $\pm$ 20% where data exists, $e^+ = e^-$, $\mu^+ = \mu^-$. Furthermore at large $p_\perp$ where comparable measurements exist we seem to have $\mu^\pm \simeq e^\pm$ as well. However, as mentioned above these equalities are not unique signatures of electromagnetic production since they will also be satisfied by the weak mechanisms except in the target and beam fragmentation regions.

A signature, unique to a weak parity violating process would be a non-zero longitudinal polarization of the direct lepton. A recent BNL/YALE/FNAL experiment[18] measured a polarization $P = 0.00 \pm 0.10$ for prompt muons produced at $p_\perp = 0$ and $x = 0.5$. An earlier Serpukhov experiment[19] observing muons at $p_\perp = 1.8$ GeV/c and $x \approx 0$ also found no evidence for longitudinal polarization. One is encouraged to believe, therefore, that a substantial part of the direct lepton signal is electromagnetic in origin.

As mentioned earlier there are many equivalent languages which may be used to discuss production mechanisms. The most popular, and one hopes fundamental, is the quark-parton language, used for example in the pioneering work of Drell and Yan.[20] Equally valid and sometimes more convenient is a description in terms of physical hadron states, such as occurs in multiperipheral[21] and vector dominance models.[22,23] Curiously, we seem to know more about the parameters and distribution functions for the unobservable quark-partons than we know about the corresponding quantities for observable hadrons. A third, much less developed language, uses proto-hadrons[24] (e.g., clusters) as the fundamental building blocks. Independent of language, however, the basic physical processes are the two familiar from QED annihilation and bremsstrahlung.[25]

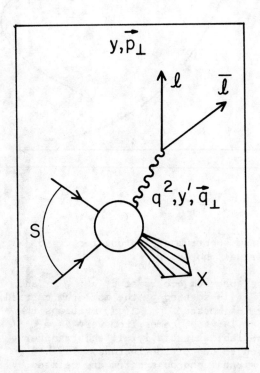

We specify the four-momentum q^μ of the virtual photon by a mass q^2, a center-of-mass rapidity y' and a transverse momentum $\vec{q}_\perp$. Similarly we specify the four-momentum p^μ of the detected lepton by a rapidity y and a transverse momentum $\vec{p}_\perp$ - see Fig. 2.

Fig. 2. Kinematical variables for lepton pair production via a virtual photon.

The spectrum in $p_\perp$ of a lepton which results from the decay of the parent virtual photon has the familiar shape shown in Fig. 3.

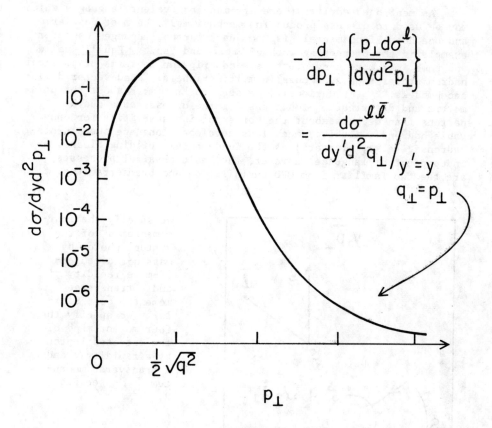

Fig. 3. The transverse momentum spectrum of leptons for y = 0 coming from the decay of a virtual photon.

The large Jacobian peak in the spectrum occurs at $p_\perp = \frac{1}{2}\sqrt{q^2}$ and comes from symmetric decays. It is smeared by the momentum carried by the virtual photon and for low masses $\sqrt{q^2} \lesssim \langle q_\perp \rangle$ overlaps the origin, $p_\perp = 0$. The spectrum at large $p_\perp$ comes from very asymmetric decays in which (in the lab) essentially all the transverse momentum of the parent is given to one of the leptons. In this region the lepton and parent virtual photon spectra are related by the Sternheimer formula[26]

$$-\frac{d}{dp_\perp}\left\{p_\perp \frac{d\sigma^\ell}{dy\,d^2 p_\perp}\right\} = \left(\frac{d\sigma^{\ell^+\ell^-}}{dy'\,d^2 q_\perp}\right)_{\substack{y'=y \\ q_\perp = p_\perp}} \qquad (2)$$

for

$$p_\perp \gg \sqrt{q^2},\ \langle q_\perp \rangle . \qquad (3)$$

It should be emphasized that both inequalities (3) must be satisfied before Eq. (2) can be employed. This is not the case for the ψ/J in the range $p_\perp \lesssim 6$ GeV studied thus far.

We can divide the models for production of large $p_\perp$ leptons into two extreme categories:

(i) High $q^2 \sim 4p_\perp^2$, low $q_\perp \sim 0$

and

(ii) Low q^2, large $q_\perp \sim p_\perp$.

HEAVY MASS PAIRS

Examples of Type (i) models are: the original Drell-Yan parton annihilation model[20] and subsequent work[27] in which no transverse momentum structure was included in the parton distributions, cluster models,[24] and models[23] which appeal to the <u>heavy vector mesons</u> of the GVDM model. All of these models are representations of the same underlying physics. The ultimate in simplicity is the ψ/J model which posits that the entire large $p_\perp$ spectrum comes from ψ/J decays.

Type (i) models, while not ruled out entirely, can easily be eliminated as the major source of prompt leptons at large $p_\perp$. The argument is basically kinematical. Consider, for definiteness, the cluster model of Pokorski and Stodolsky.[24] They picture the hadron-hadron collision as forming a cluster of mass M at rest in the center-of-mass system, where

$$\frac{d\sigma}{dM} = \beta \left(\frac{M_0}{M}\right)^\gamma \tag{4}$$

$\beta, \gamma \sim$ constant.

The cluster decays either to hadrons or electromagnetically to a $\ell^+\ell^-$ pair, $q^2 = M^2$, with a branching ratio B which they take from $B = \sigma(e^+e^- \to \mu^+\mu^-)/\sigma(e^+e^- \to \text{hadrons}) = \alpha^2 R/9$ as measured at colliding rings.[28] The single lepton spectrum at $y = 0$ is

$$\left.\frac{dN^\ell}{dy d^2 p_\perp}\right|_{y=0} = \frac{B}{4\pi p_\perp} \delta\left(p_\perp - \frac{M}{2}\right). \tag{5}$$

Hence, the dilepton spectrum is simply predicted by the single lepton spectrum

$$B \frac{d\sigma}{dM} = \frac{d\sigma^{\ell^+\ell^-}}{dM} = 2\pi \left(p_\perp \frac{d\sigma^\ell}{dyd^2p_\perp} \right)_{\substack{y=0 \\ p_\perp = \frac{M}{2}}}$$

Using as input the e^- spectrum measured (CCRS) at CERN-ISR[29] one obtains the solid curve shown in Fig. 4 along with the upper bound on $d\sigma^{\ell^+\ell^-}/dM$ for $M \geq 4$ GeV/c^2 obtained from an early (CCR) CERN-ISR pair experiment.[30] The cluster model badly violates the bound over the mass range $M = 4 - 7$ GeV/c^2 which corresponds to single leptons at $p_\perp = 2.0 - 3.5$ GeV/c.[31]

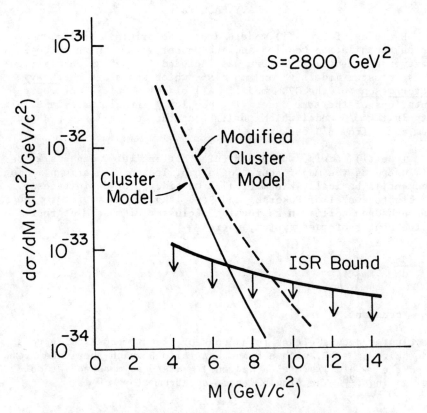

Fig. 4. The dilepton production cross section vs pair mass required to explain the single lepton spectrum observed at ISR (Ref. 29) assuming (a) the pair is at rest in the center of mass (solid curve) and (b) the pair has a uniform rapidity distribution and an exponential transverse momentum Eq. (7) (dotted curve). The experimental bound comes from Ref. 30.

The assumption that the dilepton system is a rest in the center-of-mass is obviously extreme. Assuming instead

$$\frac{d\sigma}{dMdy'd^2q_\perp} = f(M)e^{-bq_\perp}, \quad b = 4.25 \text{ (GeV/c)}^{-1} \tag{7}$$

and choosing $f(M)$ to fit the CCRS e^- spectrum one generates the dotted curve shown in Fig. 4. Again the experimental bound is violated.

Next consider the contributions from the ψ/J (3.1) and ψ' (3.7). The recent result from a COL/ILL/FNAL/HAWAII experiment[32]

$$\frac{\sigma(n+Be \to \psi'+X)B(\psi' \to \mu^+\mu^-)}{\sigma(n+Be \to \psi/J+X)B(\psi/J \to \mu^+\mu^-)} \leq \frac{1}{70} \tag{8}$$

for $x > 0.3$ eliminates the ψ' as an important source of prompt leptons at Fermilab energies. There is also no sign of the ψ' at BNL,[33] Serpukhov,[34] and CERN-ISR[35] at a level which will contribute significantly to direct lepton production. Higher mass states, ψ (4.4), etc. observed at SLAC-SPEAR[36] have normal dilepton branching ratios $B \sim \alpha^2$ and hence, even if produced as strongly as the ψ/J, will contribute negligibly to the single lepton spectrum.

This leaves the ψ/J itself as a single and unique source of high mass lepton pairs. The production of ψ/J at Fermilab energies[37] is now quite well known for all x and for transverse momenta $q_\perp \leq 2-3$ GeV/c. The transverse momentum spectrum is definitely broader than the ρ°. However, nothing is known about the rate at which ψ/J production falls off at large $q_\perp$. Experiments[35] at CERN-ISR, although less complete, indicate ψ/J production in p-p collisions at about the same rate as at Fermilab energies.

A large number of authors,[22,38] including the experimental groups themselves,[35,37] have calculated the contribution from ψ/J to the single lepton spectrum. The verdict seems to be that ψ/J production at Fermilab energies is about a factor of 2 too small to explain all of the direct leptons at $p_\perp \approx 1.5$ GeV/c but could be adjusted to fit the spectrum beyond $p_\perp \approx 2$ GeV/c given our lack of knowledge about production in the large $q_\perp$ region. However, at CERN-ISR energies the ψ/J contribution even in the Jacobian peak region is insufficient by a factor of 3-10. This comes about simply because ψ/J production increases very little from $S = 600$ GeV^2 to $S = 2800$ GeV^2 whereas the single lepton spectrum at fixed $p_\perp$ shows a substantial increase.

PAIR EXPERIMENTS

It remains, therefore to consider the contribution from low mass virtual photons — both continuum and resonant cases. Before focusing on the very low mass region let us look at the pair experiments that do exist and compare to the popular Drell-Yan[20] parton annihilation model. This is the subject that started all the excitement beginning with the pioneering low-resolution, high-acceptance BNL experiment of Christenson, et al.[39] followed by the equally pioneering high-resolution, low-acceptance BNL experiment of Aubert, et al.[40]

We now know[41] that the shoulder observed in the first BNL experiment is caused by the ψ/J smeared by the broad experimental resolution. This, along with estimates of nuclear corrections[42] for the uranium target used in the first BNL experiment, reduces the pair yield which must be explained by other sources. The remaining amount shows the general shape expected from the Drell-Yan annihilation model (including color) but is still a factor of 2 - 3 too large in magnitude - see Fig. 5 (a). In the pair mass region 4 - 5 GeV/c^2 where one is running up against the kinematic limit the disagreement is worse.[43] An analysis of the more recent high resolution[33] BNL data (S = 54 GeV^2, p + Be) which explored the 3.2 - 4.0 GeV/c^2 region would be useful, as would additional pair experiments at BNL and CERN-PS energies.

Moving up in energy we now have muon pair data[34] from Serpukhov (S = 132 GeV^2, p + Be) covering 0.5 - 3.5 GeV/c^2 in which clear $\rho + \omega$, ϕ and J/ψ peaks are observed - Fig. 5 (b). This seems to be the first time the ϕ has been seen in a dilepton experiment. No comparison of these data to resonant and continuum production models has yet been made. It should be.

At Fermilab (S = 620 GeV^2, n + Be) muon pair data[44a] in the range 1.0 - 2.4 GeV/c^2 and $q_{\parallel}^L$ > 60 GeV/c shows a magnitude and shape remarkably like the parton annihilation predictions - Fig. 5 (c). Given the factor of 2 - 3 uncertainty in the overall experimental normalization a detailed comparison is not possible. The model predictions again seem somewhat low, however. An electron pair experiment[44b] at Fermilab (S = 750 GeV^2, p + Be) has recently reported results for the high mass range 5 - 11 GeV/c^2. As we heard earlier this morning in J. Appel's talk the magnitude and shape is in remarkable agreement with the Drell-Yan model calculated with a factor of $\frac{1}{3}$ for color and reasonable parton model structure functions - Fig. 5 (d).

I might remark that whereas there has been reluctance in the past to accept the factor of 3 reduction that color implies for the Drell-Yan model, it seems inescapable if we are to remain faithful to the rules of quantum mechanics. Color has become so central to our thinking about quark binding and confinement that no theorist, at least, would want to do without it.

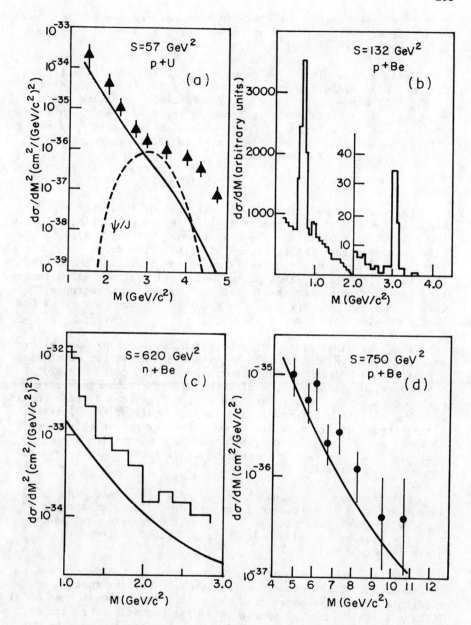

Fig. 5. Comparison of Drell-Yan model (solid curves) and experiment: (a) BNL data (Ref. 39) with nuclear corrections (Ref. 42); (b) Serpukhov data (Ref. 34); (c) Fermilab data (Ref. 44a); (d) Fermilab data (Ref. 44b). Structure functions are taken from Ref. 45.

Pair experiments[35] at CERN-ISR have not yet reached the level of sensitivity that cross sections of the size predicted by the parton annihilation model could be detected.

Taken as a whole the evidence seems quite impressive for the presence of a virtual photon continuum 1 - 10 GeV/c^2 in addition to the well-known ρ, ω, ϕ and J/ψ resonances. The shape and magnitude is understandable in terms of the Drell-Yan annihilation model except for a factor of $2 \sim 3$ excess in the lower mass region. This continuum can also be interpreted in the language of the generalized vector dominance model.[23]

LIGHT MASS PAIRS

Earlier we showed that if virtual photons (pairs) are produced at low $q_\perp$ only, one cannot explain the large $p_\perp$ single lepton spectrum without violating the experimental value for $d\sigma/dq^2$. We have just seen that the experimental results for $d\sigma/dq^2$ are in qualitative agreement with the Drell-Yan parton annihilation model. It follows, therefore, that parton model calculations which ignore the transverse momentum content of the partons in a hadron will fail to explain the single lepton spectrum at large $p_\perp$. This has been known for some time.

Thus the usual parton structure functions $q(x)dx$ must be generalized to $q(x,\vec{k}_\perp)dx\, d^2k_\perp$ which gives the probability of finding a parton in a hadron with longitudinal momentum faction x and transverse momentum $\vec{k}_\perp$. The distribution functions $q(x,\vec{k}_\perp)$ are presumed to be strongly peaked about $k_\perp = 0$ so that Bjorken scaling for νW_2, etc. is unaffected. The large $k_\perp$ tail of the distributions, though small, provides a mechanism for producing virtual photons, and thereby leptons, at large $p_\perp$. Kinematics and the sharp fall of $d\sigma/dq^2$ with increasing q^2 dictates that leptons at large $p_\perp$ come from low q^2 photons.

Calculations[46] show the expected improvement. The single lepton spectrum is increased in magnitude at large $p_\perp$ and has a shape which much more closely resembles the experimental spectrum than the shape from naive calculations which correspond to $q(x,\vec{k}_\perp) = q(x)\delta(\vec{k}_\perp)$.

Detailed comparison depends, of course, on the details of the $k_\perp$ dependence of the structure functions. The only insight we have into this at present comes from the constitutent interchange model (CIM)[47] in which the $k_\perp$ dependence is calibrated by the rate of hadron production at large $p_\perp$. This is the approach used in work cited above.

In the CIM scheme one has a host of subprocesses[48] in addition to the Drell-Yan parton annihilation process, each one having its own particular q^2 and $q_\perp$ dependencies. One can anticipate that this subject be an active one in the near future.[49]

In detailed comparisons to single lepton spectra one question needs to be faced. Should one subtract the contribution from ψ/J decay before comparing to theoretical models or not? It is common to finesse this question by appealing to duality. However, one can equally well argue that the ψ/J is important only because of an accident - its anomalously small hadronic width and corresponding anomalously large branching ratio into $\ell^+\ell^-$. I have not yet seen a convincing demonstration that ψ/J production is subsumed in the usual parton model calculations.

In addition to the above work on transverse momentum structure, there has been a considerable body of work devoted to seeking corrections and additions to the Drell-Yan process. One such mechanism, proposed long ago is bremsstrahlung.[50] The issue has been reraised recently by Farrar and Frautschi[51] who emphasized the possibility of copious production of direct (real) photons at large $p_\perp$. Such a situation would complicate the π° and η measurements and the subsequent subtraction of their Dalitz decays when seeking the direct lepton spectrum. Real photons ought to be accompanied by (predominately low mass) virtual photons produced in the same way.

While the qualitative features of a bremsstrahlung mechanism: large peak at low q^2 ($d\sigma \propto 1/q^2$) and a rapid fall off at high mass ($d\sigma \sim \exp(\beta\sqrt{q^2})$) are well known, detailed calculations of the amount expected are difficult and non-existent. The issue will have to be settled experimentally by measurements of real photons at large $p_\perp$ and further comparisons of the e/μ ratio. One cannot help but recall the excess of real photons at large $p_\perp$ observed in the UCSB/SLAC inelastic Compton experiment.[52]

The annihilation mechanism may be modified at low q^2 as well. Because it is not controlled by light - cone singularities,[53] the Drell-Yan process has never enjoyed the prestige of deep inelastic scattering. Indeed corrections have been suggested[54] and debated[55] in the literature. The general feeling seems to be that in the large q^2 region asymptotic freedom or its equivalent may be operative suggesting that the Drell-Yan formula should be unmodified. On the contrary for low q^2, modifications are not unreasonable even if we are not able to calculate them well. An interesting attempt along this line has been made by Bjorken and Weisberg.[56] They suppose that the $q\bar{q}$ pair which annihilate to form the virtual photon come not from quarks which are present initially in the beam and target wave functions but are instead newborn quarks which have not yet found their way into the secondary particles $(\pi, K, ...)$ which populate the final states of high energy hadron-hadron collisions. [Presumably this is somewhere contained in the general framework laid out in Ref. (48).] Using the yield of π's as a guide they estimate an enhancement of ~ 25 over the usual Drell-Yan formula for low masses. Needless to say such an enhancement is an attractive option. It would be nice to see a space-time analysis of their picture.

It is generally agreed that for low masses the quark-parton language loses its power and clarity. Simple vector meson dominance becomes the language of choice. By using as input the experimental valves for ρ, ω, ϕ, etc. production one bypasses the complicated production mechanism issue. We all know, however, that the results are troublesome. Calculations[22,38] based on measured parameters for the light vector mesons supplemented with reasonable guesses where data are absent fail to explain the single lepton spectrum. The problem is acute in the region $p_\perp < 1.0$ GeV/c where the uncertainty, discussed above, of the transverse momentum spectrum of the vector meson itself is not important. If in fact there is a large non-resonant continuum at low masses which dominates the resonant parts, it would imply a massive breakdown of simple vector dominance. This would be hard to believe and I, at least, would not welcome it.

Two experimental groups[57] this morning presented evidence for a broad pair spectrum concentrated in the region $\sim 0.5 - 2.5$ GeV/c^2 which they argue can explain the integrated single lepton spectrum. It will be interesting to see detailed analysis of their mass distributions.

SUMMARY

We have seen:

(1) There is growing evidence for the production of a virtual photon continuum which is remarkably like the Drell-Yan model predictions.

(2) Single lepton spectra at large $p_\perp$ comes predominantly from low q^2, large $q_\perp$ parents.

(3) The parton annihilation model calculations at low q^2 probably needs to be enhanced by one or more mechanisms.

(4) The contribution of charmed particle decay to the direct lepton spectrum is unknown at present.

(5) Pair experiments will ultimately settle the issues now before us. We welcome them.

REFERENCES

1. The situation as of August 1975 is reviewed by L. Lederman, <u>Proceedings of the 1975 International Symposium of Lepton and Photon Interactions at High Energies</u> edited by W. T. Kirk (Stanford Linear Accelerator Center, Stanford, CA, 1975) p. 265.

2. See L. Leipuner, I. Gaines, J. Appel, M. Chen, C. Rubbia, G. Sanders, H. Lubatti, R. Ruchti, and M. Mallary, these Proceedings.

3. T. D. Lee and G. C. Wick, Phys. Rev. D1, 1033 (1970).

4. R. Burns, et al., Phys. Rev. Lett. 15, 830 (1965); R. C. Lamb, et al. Phys. Rev. Lett. 15, 800 (1965); P. J. Wanderer, et al. Phys. Rev. Lett. 23, 729 (1969).

5. S. Weinberg, Phys. Rev. Lett. 19, 1264 (1967); A. Salam, <u>Elementary Particle Theory: Relativistic Groups and Analyticity</u> (Nobel Symposium No. 8) edited by N. Svartholm (Amquist and Wiksel, Stockholm, 1968), p. 367.

6. S. L. Glashow, J. Iliopoulos, and L. Maiani, Phys. Rev. D2, 1285 (1970). For a review see M. K. Gaillard, B. W. Lee, and J. L. Rosner, Rev. Mod. Phys. 47, 277 (1975).

7. For a review see R. M. Barnett, Phys. Rev. D13, 671 (1976) and references therein.

8. T. Applequist, A. De Rújula, H. D. Politzer, and S. L. Glashow, Phys. Rev. Lett., 34, 365 (1975); E. Eichten, K. Gottfried, T. Kinoshita, J. Kogut, K. D. Lane and T.-M. Yan, Phys. Rev. Lett. 34, 369 (1975).

9. J.-E. Augustin, et al., Phys. Rev. Lett. 34, 764 (1975); J. Siegrist, et al., Phys. Rev. Lett. 36, 700 (1976).

10. See, for example: T. Ferbel, Ann. N.Y. Acad. Sci. 229, 124 (1974); T. K. Gaisser and F. Halzen, Phys. Rev. D11, 3157 (1975); D. Sivers, Phys. Rev. D11, 3253 (1975); L. Pilachowski and S. F. Tuan, Phys. Rev. D11, 3148 (1975); R. D. Field and C. Quigg, Fermilab Report No. 75/15 - THY (unpublished).

11. M. B. Einhorn and S. D. Ellis, Phys. Rev. Lett. 34, 1190 (1975); Phys. Rev. D12, 2007 (1975).

12. M. K. Gaillard, B. W. Lee and J. L. Rosner, Rev. Mod. Phys. 47, 277 (1975). For an early view see G. A. Snow, Nucl. Phys. B55, 445 (1973). Recent work: G. Altarelli, N. Cabibbo, and L. Maiani, Nucl. Phys. B88, 285 (1955); M. B. Einhorn and C. Quigg, Phys. Rev. D12, 2015 (1975); R. L. Kingsely, S. B. Treiman, F. Wilczek, and A. Zee, Phys. Rev. D11, 1919 (1975); I. Karliner, Phys. Rev. Lett. 36, 759 (1976); J. Ellis, M. K. Gaillard, and D. V. Nanopoulos, Report No. TH 2030-CERN, 1975 (to be published).

13. A. M. Boyarski, et al., Phys. Rev. Lett. 35, 196 (1975). See also the theoretical comments in M. B. Einhorn and C. Quigg, Phys. Rev. Letters 35, 1114 (1975).

14. G. Altarelli, N. Cabibbo and L. Maiani, Phys. Rev. Lett. 35, 635 (1975). One of the motivations for this work was an explanation in terms of charmed particle production of the SPEAR events (Ref. 15) $e^+e^- \to e + \mu +$ undetected neutrals for which the most attractive interpretation is the pair production of new, heavy leptons. For a critique of the charmed particle interpretation in terms of recent SPEAR data see G. Snow, Phys. Rev. Lett. 36, 766 (1976).

15. M. Perl, et al., Phys. Rev. Lett. 35, 1489 (1975).

16. L. Baum, et al., Phys. Lett. 60B, 485 (1976); C. Rubbia, these Proceedings.

17. L. M. Lederman and S. White, Phys. Rev. Lett. 35, 1543 (1976).

18. L. B. Leipuner, et al., BNL/YALE/FNAL preprint (1976).

19. V. V Abramov, et al., Proceedings of the XVII International Conference on High Energy Physics, edited by J. R. Smith (Science Research Council, Rutherford Laboratory, Chilton, Didcot, England. 1974), p. V-53, (presented by S. Nurushev).

20. S. D. Drell and T.-M. Yan, Phys. Rev. Lett. 25, 316 (1970); Ann. Phys. 66, 578 (1971).

21. H. D. I. Abarbanal and J. B. Kogut, Phys. Rev. D5, 2050 (1972); G. Chiu and J. Koplik, Phys. Lett. 55B, 466 (1975); Phys. Rev. D11, 3134 (1975); K. Subbarao, Phys. Rev. D8, 1498 (1973); H. B. Thacker, Phys. Rev. D2, 154 (1970).

22. M. Bourquin and J. -M. Gaillard, Phys. Lett. 59B, 191 (1975); F. Halzen and W. F. Long, Univ. of Wisc. preprint COO-480 (1975) (to be published).

23. J. J. Sakurai and H. B. Thacker, Nucl. Phys. B76, 445 (1974); F. M. Renard, Nuovo Cimento 29A, 64 (1975); M. Greco and Y. N. Srivastava, Nucl. Phys. B64, 531 (1973); E. Etim, M. Greco, and Y. N. Srivastava, Phys. Lett 41B, 507 (1972); S. Chavin and J. D. Sullivan, Phys. Rev. (to be published); E. N. Argyres, A. P. Contogouris and C. S. Lam, McGill Univ. Preprint (1976); A. P. Contogouris and A. Nicolaidis, McGill Univ. Preprint (1976); F. Halzen, Univ. of Wisc. Preprint COO-501 (1976).

24. S. Pokorski, and L. Stodolsky, Phys. Lett. 60B, 84 (1975); L. Montvay, Phys. Lett. 53B, 377 (1974).

25. For an early, but nonetheless still valid discussion of these issues see: R. L. Jaffe, Ann. N.Y. Acad. Sci. 229, 225 (1974).

26. R. M. Sternheimer, Phys. Rev. 99, 277 (1955). For a treatment in terms of modern variables see R. N. Cahn, Phys. Rev. D7, 247 (1973).

27. See for example: S. M. Berman, J. D. Bjorken and J. B. Kogut, Phys. Rev. D4, 3388 (1971); P. V. Landshoff and J. C. Polkinghorne, Nucl. Phys. B33, 221 (1971); B36, 642 (E) (1972); G. R. Farrar, Nucl. Phys. B77, 429 (1974); F. Renard, Nuovo Cimento 29A, 64 (1975); R. McElhaney and S. F. Tuan, Phys. Rev. D8, 2267 (1973); Nucl. Phys. B72, 487 (1974); H. Paar and E. Paschos, Phys. Rev. D10, 1502 (1974); S. Pakvasa, D. Parashar, and S. F. Tuan, Phys. Rev. D11, 214 (1975).

28. Note that this model and the closely related quark-parton model of Y. Eylon and Y. Zarmi, Phys. Lett. 56B, 47 (1975), has the feature that direct lepton production is proportional to R rather than R^{-1} as in the usual Drell-Yan model. Such a behavior is reasonable in the resonance region but is inconsistent in regions where quarks may be treated as free.

29. F. W. Busser, et al., Phys. Lett. 53B, 212 (1974).

30. F. W. Busser, et al., Phys. Lett. 48B, 377 (1974).

31. The CCR bound, because it comes from seeing zero events above background, is surely an overestimate by orders of magnitude of the pair cross section for heavy masses $M > 10$ GeV/c^2. It seems unreasonable, therefore, to believe the cluster model might be relevant for $p_\perp > 3.5$ GeV/c even though we cannot prove from the ISR bound that it is wrong.

32. See the talk by I. Gaines, these Proceedings.

33. J. J. Aubert, et al., Phys. Rev. Lett. 33, 1624 (1974).

34. Y. M. Antipov, et al., Phys. Lett., 60B, 309 (1976).

35. F. W. Busser, et al., Phys. Lett., 56B, 482 (1975); E. Nagy, et al., Phys. Lett., 60B, 96 (1975).

36. J. Siegrist, et al., Phys. Rev. Lett. 36, 700 (1976); J.-E. Augustin, et al., Phys. Rev. Lett. 34, 764 (1975).

37. B. Knapp, et al., Phys. Rev. Lett. 34, 1044 (1975); K. J. Anderson, et al., Phys. Rev. Lett. 36, 237 (1976); G. J. Blanar, et. al., Phys. Rev. Lett. 35, 346 (1975).

38. F. Halzen and K. Kajantie, Phys. Lett. 57B, 361 (1975); G. R. Farrar and R. D. Field, Phys. Lett. 58B, 180 (1975); J.-M. Gaillard, Proceedings of the Recontre de Moriond, Meribel-les-Allues, France (1975); J. A. Appel, these Proceedings; S. Chavin and J. D. Sullivan, Ref. 23; E. N. Argyres, A. P. Contogouris and C. S. Lam, Ref. 23; A. P. Contogouris and A. Nicolaidis, Ref. 23.

39. J. H. Christenson, et al., Phys. Rev. Lett. $\underline{25}$, 1523 (1970).

40. J. J. Aubert, et al., Phys. Rev. Lett. $\underline{33}$, 1404 (1974).

41. Y. Eylon and Y. Zarmi, Phys. Lett. $\underline{56B}$, 47 (1975); M. Duong-van, preprint, SLAC-PUB-1603 (1975).

42. G. R. Farrar, preprint, CALT 68-497 (1975).

43. For a useful discussion in terms of bounds see M. B. Einhorn and R. Savit, Phys. Rev. Lett. $\underline{33}$, 392 (1974), Phys. Rev. $\underline{D10}$, 2781 (1974).

44a. T. O'Halloran, Proceedings of the 1975 International Symposium on Lepton and Photon Interactions at High Energies, edited by W. T. Kirk (Stanford Linear Accelerator Center, Stanford, CA 1975), p. 189. There is perhaps evidence for a ϕ peak in this experiment as well.

44b. D. C. Hom, et al., Columbia/Fermilab/Stony Brook preprint (1976); J. Appel, these Proceedings.

45. G. Chu and J. F. Gunion, Phys. Rev. $\underline{D10}$, 3672 (1974).

46. M. Duong-van, SLAC-PUB-1604 (1975) (to be published) C. O. Escobar, preprint DAMTP 75/9 (1975) (to be published); J. F. Gunion, Davis preprint (1976) (to be published).

47. For a review see: D. Sivers, S. J. Brodsky and R. Blankenbecler, Phys. Reports $\underline{23C}$ (1976).

48. C. T. Sachrajda and R. Blankenbecler, SLAC-PUB-1594 (1975) (to be published); SLAC-PUB-1658 (1975) (to be published).

49. The subprocess $M+q \rightarrow \gamma^* +q$ has been studied by M. Fontannaz, LPTHE (Orsay) 75/29 (1975) (to be published).

50. S. M. Berman, D. J. Levy and T. L. Neff, Phys. Rev. Lett. $\underline{23}$, 1363 (1969); G. Altarelli, R. A. Brandt and G. Preparata, Phys. Rev. Lett. $\underline{26}$, 42 (1971); R. A. Brandt and G. Preparata Phys. Rev. $\underline{D6}$, 619 (1972).

51. G. R. Farrar and S. C. Frantschi, CALT-68-518 (1975) (to be published).

52. D. O. Caldwell, et al., Phys. Rev. Lett. $\underline{33}$, 868 (1974).

53. R. L. Jaffe, Phys. Lett. $\underline{B37}$, 517 (1971). See also the discussion in K. Wilson, Proceedings of the 1971 International Symposium on Photon Interactions at High Energies, edited by N. B. Mistry (Cornell University, Ithaca, N.Y. 1972), p. 115.

54. P. V. Landshoff and J. C. Polkinghorne, Ref. 27.

55. F. Henyey and R. Savit, Phys. Lett. $\underline{52B}$, 71(1974); J. L. Cardy and G. A. Winbow, Phys. Lett. $\underline{52B}$, 95 (1974); M. B. Einhorn and F. S. Henyey, Phys. Rev. $\underline{D11}$, 2009 (1975); C. E. DeTar, S. D. Ellis, and P. V. Landshoff, Report No. TH 1925-CERN (1974) (to be published).

56. J. D. Bjorken and H. Weisberg, SLAC-PUB-1631 (1975) (to be published).

57. Chicago/Princeton, G. Sanders, these proceedings; Yale/BNL/Fermilab, L. B. Leipuner, these Proceedings.

7 FT. BUBBLE CHAMBER ν EXPERIMENT
at
BROOKHAVEN NATIONAL LABORATORY

E. G. Cazzoli

A. M. Cnops, P. L. Connolly, R. I. Louttit, M. J. Murtagh
R. B. Palmer, N. P. Samios, T. T. Tso, H. H. Williams[*]
Brookhaven National Laboratory, Upton, L. I., New York 11973

ABSTRACT

In this talk are presented preliminary results on the 7' Bubble Chamber ν Experiment concerning exposures in H_2 and D_2, with a yield of 850 ν interactions; in particular we analyze 10 charged current strange particle events, that provide a cross section ratio to non strange particle interactions of $.04 \pm .02$ at our energies.

INTRODUCTION

The AGS beam of around 5×10^{12} protons per pulse with a momentum of 8 GeV/c is ejected, meets a target and charged π's and K's are focused towards the 7' Bubble Chamber by a double horn system; at the chamber the vertical spread of the ν beam from the decay of above mentioned particles is around ± 1m and the total area is around $4m^2$. The Bubble Chamber itself is a cylinder of 2.1m in diameter and around 3m high. The visible volume is $6m^3$ and the fiducial volume around $4m^3$. The reduction is mainly due to 4 steel plates, 2" thick each, set inside the chamber to aid in separation of leptons and strongly interacting particles, and in conversion of γ's. The chamber normally operates at 1.5 sec. repetition rate in a 25 KGauss magnetic field (superconductor).[1]

ANALYSIS

Up to date we had 3 exposures in H_2 and 3 in D_2, with a yield of pictures and events as shown in Table I.

Exp.	Pict. x 10³	ν Events	Analyzed
H_2	90	130	Yes
D_2	350	780	Yes
H_2	130	300	No

TABLE I

[*]Present Address: University of Pennsylvania, Philadelphia, Pennsylvania 19104

We expect to have another exposure in D_2; furthermore a collaboration Columbia-BNL is expected to run with a Neon-H_2 mixture. We have under study improvements on the horn system which should double the flux at higher energies.

The film has been all double scanned and the scanning efficiency has been found to be a function of the number of charged particles N in the final state, rising from .8 to 1. as N goes from 1 to ≥ 5. Charged tracks are thus identified:

1) μ^-, π^-: the plates, interactions and decays allow a separation with around 60% efficiency

2) π^+, p: positively charged tracks are separated up to 85% of the time by stopping, decaying, interacting, overstopping, ionization criteria.

For neutral particles, mostly π^o and neutrons, the plates and liquid together provide a 40% probability of conversion or interaction.

SOURCES OF BACKGROUND

All the interactions occurring in the chamber have been measured. An **obvious** source of background -- elastic and inelastic scattering of charged particles -- is rejected by a check on the coplanarity of the event and with a cut in the total visible momentum at 400 MeV/c.
In Fig. 1 is shown the visible direction of all the events that are left. It is apparent that events with a visible dip of more than 30° can be ascribed to cosmic ray neutrons interactions.
We can thus define as a neutrino induced reaction any event with a polar angle with respect to incoming ν direction $\theta_{vis} \leq 40°$.

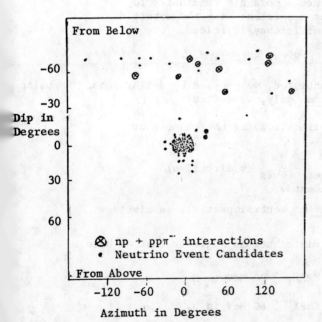

Azimuth versus Dip of the Total Momentum Vector for all Candidates for Neutrino Interactions in Hydrogen. The Nominal Beam Direction is Azimuth of Zero, Dip of Zero and Vectors Pointing Down Have Positive Dips.

FIG. 1

As for background in the general beam direction

a) neutrons: we accept as normalization the 1C reaction $n_p \to \pi^- pp$, at present ≤ 4 events. Using Form. 1,

1) $$\frac{N(\text{n interactions in a given channel})}{\phi(E) = N_o e^{-4E}} = \frac{\int_{E_{min}}^{E_{max}} \phi(E)\sigma(E) \Big|_{\text{Channel}} dE}{\int_{E_{min}}^{E_{max}} \phi(E)\sigma(E) \Big|_{pp\pi^-} dE}$$

we can claim that the total neutron background is ≤ 28 events.

b) K_L^o from neutrino interactions in the magnet coil can be normalized to the neutron background at ≤ 1. event.

c) $\bar{\nu}$ contamination: the expected flux of $\bar{\nu}$ in our beam is .03; folding in the cross section for $\bar{\nu}$ interaction at our energies, the total background is expected to be $\sim 1\%$. Analysis of the events showed that there are 3 possible candidates for $\bar{\nu}N \to \mu^+ + X$, number consistent with the expected, once μ^+ detection efficiency is folded.

EVENT SELECTION

To keep the selection unbiased, we proceed by definitions. Focusing on <u>charged current interactions</u> only, we define:

1) a μ^- is a negative track leaving the chamber or decaying to an e^-

2) EMPX = $E_{vis} - m_{target} - P_{vis}$ (ν direction) is the mass of the incident ν

it is $0 \pm \begin{matrix} 20 \text{ MeV in } H_2 \\ 80 \text{ MeV in } D_2 \end{matrix}$ <u>if</u> no neutral particle is missing.
(See Fig. 2)
A further study of the distributions suggests that we have a possible 3C Fit if

$\theta_{vis, \nu} \leq 4°$ $|EMPX| \leq 20$ MeV in H_2

$\theta_{vis, \nu} \leq 6°$ $|EMPX| \leq 80$ MeV in D_2

Outside these cuts the selection is done by looking at the events on the table - the $\sim 15\%$ unresolved ambiguous events are then properly weighted.

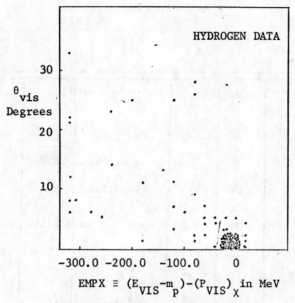

Plot of EMPX versus the Angle of the Visible Momentum Vector with Respect to the Beam Direction.

FIG. 2

CHARGED CURRENT STRANGE PARTICLE PRODUCTION

Up to date we observed 10 events that checked the selection criteria (Table 2), and for the present analysis we used 6 of them. An important point to note is that we can rely on 3C fits for associate production identification, thus being able to separate between $\Delta S = \pm \Delta Q$ and $\Delta S = 0$ channels.

The raw number of events has to be corrected (1) by decay rates[2] (2) by momentum and position detection probability (Montecarlo generation). The corrected numbers thus are:

$$\begin{array}{r} 5.1 \pm 3.6 \text{ in } H_2 \\ \underline{8.6 \pm 4.3} \text{ in } D_2 \\ 13.7 \pm 5.6 \text{ Total} \end{array}$$

CHARGED CURRENT
STRANGE PARTICLE EVENTS

REACTION	ID	E_ν	TYPE	W_{HAD}	Q^2	X	Y
$\nu p \to \mu^- \Lambda^o \pi^+ \pi^+ \pi^+ \pi^-$	3C	13.5	$\Delta S = -\Delta Q$	2.426 ±.013	2.17	.31	.28
$\nu p \to \mu^- K_S^o \pi^+ p$	3C	5.2	$\Delta S = \pm\Delta Q$	2.059	.47	.12	.39
$\nu n \to \mu^- K^+ \pi^- \Sigma^+ p_S$	3C	7.	Assoc.Prod.	2.086	6.5	.72	.69
$\nu n \to \mu^- \pi^- \pi^+ K_S^o X^+ X^o$	0C	>5	Assoc.Prod.				
$\nu n \to \mu^- \Lambda^o K^+ p_S$	3C	3.1	Assoc.Prod.	2.116	.79	.16	.87
$\nu n \to \mu^- \pi^+ (\Lambda^o) K^o / K^+ \pi^o$	0C	5.8 6.5	Assoc.Prod.	2.534 2.762	1.312 1.478	.19 .18	.64 .68
$\nu p \to \mu^- \pi^+ \pi^+ K_L^o (\Lambda^o)(n)$	0C	5.6	Assoc.Prod. ($\Delta S = -\Delta Q$)	2.873 (2.7)	.79	.1	.67
$\nu p \to \mu^- \pi^+ K^+ \Lambda^o$	3C	9.0	Assoc.Prod.	3.265	1.8	.16	.68
$\nu p \to \mu^- \pi^+ K^+ \Lambda^o$	3C	4.5	Assoc.Prod.	2.477	.39	.07	.69
$\nu p \to \mu^- \pi^+ K^+ \Lambda^o$	3C	3.4	Assoc.Prod.	2.278	1.03	.20	.79

TABLE 2

It is clear that, since more than half of our sample consists of quasi-elastic interactions, we have to choose a suitable cut to select ν interactions that are "comparable" to the strange particle ones.

We selected a cut in $\nu = E_\nu - E_\mu$ at .9 GeV, $E_\nu \geq 1$ GeV; the sample is reduced to 296 ± 15 events. An estimate of quasi elastic background is ≤ 8%, but this is comparable to losses of inelastic due to the cuts.

Thus we can compute the following ratios:

2) $R_1 = \dfrac{V^o \text{ prod.}}{\text{Inel. } \nu \text{ Inter.}} = .04 \pm .02$

3) $R_2 = \dfrac{\Delta S = \Delta Q}{\text{Inel.}} = .01 \pm .01$

4) $R_3 = \dfrac{\Delta S = -\Delta Q}{\text{Inel.}} = .006 \pm .006$

5) $R_4 = \dfrac{\text{Assoc. Prod.}}{\text{Inel.}} = .02 \pm .01$

Note that 3) is a lower limit since we did not yet include in the analysis events of the type $\nu N \to K^+ +$ non strange.

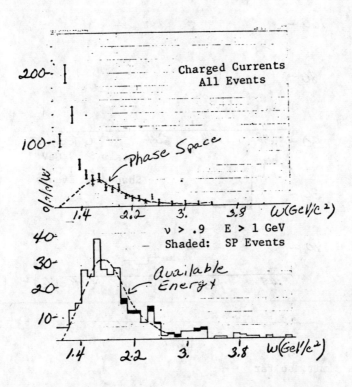

FIG. 3

In order to check whether our strange particle events are of more interest than mere "numerology" we studied all of them together with the inelastic interactions; Figs. 3, 4 and 5 show W, Q^2, x and y distributions for all events and for the selected ones; it is evident that inelastic and strange particle events show only a phase space distribution, without any strong evidence for production of high mass N^* or Σ^*, and no anomaly can be detected. Only the event shown in Fig. 6 is "different" insofar that it fits the 3C reaction

6) $\nu p \to \mu^- \Lambda^o \pi^+ \pi^+ \pi^+ \pi^-$

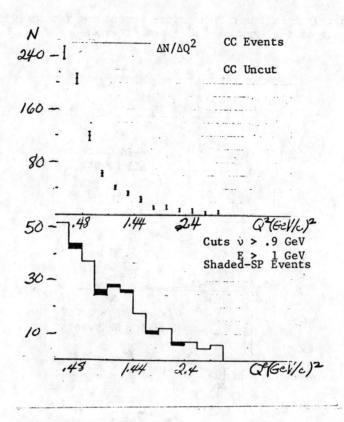

FIG. 4

thus giving a limit, so far,

$$\frac{\Delta S = -\Delta Q}{\Delta S = +\Delta Q} = .5$$

(previously quoted from $\Sigma^{\pm}$ decay$_3$ at $\sim 10^{-6}$) we have no reason to disbelieve our published arguments, also we can now add our own $p_{\perp}^2$ hadronic $p_{\perp}^2(\mu)$ distributions, that are consistent with electroproduction and other ν experiment:

7) $N(p_{\perp}^2) \mid \text{Hadron} = N_o e^{-(7\pm2) p^2}$

$N(p_{\perp}^2) \mid \mu \cong C$

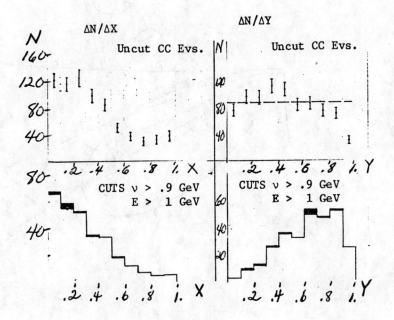

FIG. 5

The hadronic mass in interaction (6) is 2426 ± 13 MeV/c^2, thus we checked if a recurrence of this mass was apparent in all possible combinations in the inelastic sample. The only result, aside from a statistical fluctuation, is that within the products there is $\Sigma^*(1340)$ and may be $K^*(890)$.

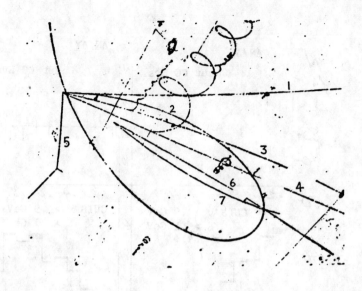

FIG. 6

COMPARISON WITH OTHER EXPERIMENTS (4), (5), (6), (7)

It is not easy to compare our result with most of the former published data, since different cuts have been applied; these are some of the numbers you can find in literature:

1) Gargamelle: cut E > 1 GeV

$$\frac{\text{Strange Particles}}{\text{All}} = 2.5\%$$

$$\frac{\Delta S = \Delta Q}{\text{All}} = .5\%$$

$$\frac{\text{Assoc. Prod.}}{\text{All}} = 2\%$$

2) ANL 12 Foot Bubble Chamber Cut $2 \leq E \leq 3$ GeV

$$\frac{\mu^- K p}{\mu^- N(n\pi)} = (4 \pm 3)\%$$

3) FNAL 15 Foot Bubble Chamber Cut $\Sigma p_x^{vis} > 10$ GeV

$$\frac{V^0}{\text{All}} = (16 \pm 2)\%$$

Fig. 7 shows $\sigma(V^0)/\sigma(\text{Tot})$ as a function of recoiling hadron mass. FNAL's and our result agree with previous electro and photoproduction experiments, showing that $\sigma(V^0)$ is increasing with energy with respect to the total cross section.

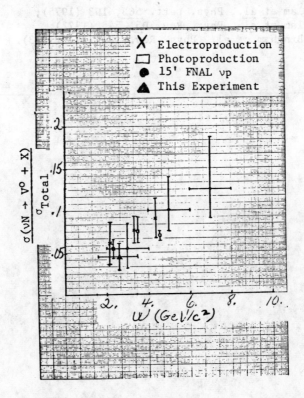

FIG. 7

REFERENCES

1. For more details on the apparatus and analysis see M. J. Murtagh in Proceedings of International Colloquium on Neutrino Physics at High Energy, Paris, March 1-3, 1975.

2. NN and ND Interactions; KN Interactions - A Compilation; Particle Data Group; UCRL-20000NN.

3. E. G. Cazzoli et al., Phys. Rev. Lett. 34, 1125 (1975).

4. M. Hagenauer in Proceedings of XVII International Conference on High Energy Physics, London, 1974.

5. S. J. Barish et al., Strange Particle Production in Neutrino Interactions, ANL/HEP 7450.

6. J. P. Berge et al., Inclusive Strange-Particle Production by νp Interactions in the 10-200 GeV Region, Phys. Rev. Lett. 36, (1976) 127 foll.

7. J. Ballam et al., Phys. Lett. 56B, 193 (1975);
J. Ballam et al., Phys. Rev. D5, 545 (1972);
H. H. Bingham et al., Phys. Rev. D8, 1277 (1973).

OBSERVATION OF THE REACTION $\nu p \to \nu p$*

C. Y. Pang
University of Illinois at Urbana-Champaign, Urbana, Illinois 61801

ABSTRACT

We have observed the reaction $\nu p \to \nu p$ in an experiment conducted at the Brookhaven AGS. Experimental details and a comparison with quasi-elastic neutrino scattering are presented.

In this talk I will discuss the observation of elastic neutrino proton scattering

$$\nu p \to \nu p \qquad (1)$$

by the Columbia-Illinois-Rockefeller Collaboration in an experiment conducted at Brookhaven National Laboratory[1]. The existence of neutral currents in neutrino interactions has been established by several experiments[2]. At the present time, however, only an upper limit has been established for elastic neutrino-proton scattering[3]. Elastic neutrino proton scattering is the simplest of the neutrino hadronic reactions and ultimately may provide a test for the theoretical models which have been proposed for neutral current interactions[4,5].

The neutrino beam was operated in the fast extraction mode with each beam spill consisting of 12 bunches which were about 35 ns wide and with a separation of 220 ns. The detector is located 150 ft. downstream from the main iron shield. There is an additional 8 ft. thick iron shield located 50 ft. upstream of the detector. The detector is an array of 21 modules of 6x6 ft. narrow gap aluminum spark chambers interspersed with scintillation counters and 8x8 ft. range chambers.

The results of this report are based on the analysis of 120,000 pictures. The film was scanned to select events consisting of one and two prongs which were consistent with the reaction (1) and the quasi-elastic reaction

$$\nu n \to \mu^- p \qquad (2)$$

with a minimum visible track range of 2 inches of Aluminum for the proton. Candidates were measured and reconstructed. We accept events with the vertex position within a 4x4 ft. fiducial area. The reconstructed events are then matched to the counter hits associated with that event so that the time-of-flight and pulse height information can be obtained.

In our analysis, a muon is defined as either a straight track with range greater than two interaction lengths, a straight track

---

*Work supported in part by ERDA.

exiting the detectors, or a stopping track with visible multiple scattering. A quasi-elastic reaction is defined as an event with one muon and one stopping straight track with range less than 20 in. of aluminum. An elastic reaction event is defined to have only one stopping straight track with range less than 20 in. of aluminum.

The target is composed of aluminum and carbon nuclei instead of free nucleons. In order to minimize the nuclear effects, such as Fermi motion of the nucleons, we require the range of the emerging proton be greater than 3 inches of aluminum, which corresponds to 550 MeV/c proton momentum. Figure 1 shows that in most of the events of reaction (2), the muons and protons tend to be coplanar as expected. We have also compared the neutrino spectrum inferred by the events satisfying the criteria for reaction (2) and find that it is consistent with the calculated neutrino spectrum.

In order to establish the existence of elastic neutrino scattering we must eliminate three major sources of background:

1. Events due to neutron interactions,

2. Single pion production by neutrinos which satisfy the scanning criteria,

3. Wide angle muons from quasi-elastic scattering in which the muon is not observed.

The experiment has several advantages in eliminating neutron interactions. Plexiglass Cherenkov detectors were inserted into the iron shield and used to start the time-of-flight clock and open the system gate upon arrival of muons from the target. The pulse height and time-of-flight were recorded for each hit in each counter. The time-of-flight (TOF) information enables us to differentiate the neutrino induced events (prompt) from the neutron induced events (delayed). In order to reduce accidental hits, we require that at least two counters along the stopping straight track are hit and we reject events which have upstream associated events in the same RF bucket. Figure 2a shows the TOF distribution for events which satisfy the elastic reaction criteria. It is apparent that there are significant off-time background events as compared to the TOF quasi-elastic data shown in Figure 2c. A significant source of these events is a neutron flux entering the detector from the top. In order to eliminate this background, we make a correlated cut on the dip angle and the height of the events in the detector. It should be noted that the detector is 7 feet off the ground and we do not observe any enhancement in the bottom half of the detector. We reduce the background further by requiring that the angle of the stopping track with respect to the neutrino direction be greater than 25°. This angular cut not only reduces the neutron background, but it also reduces the wide angle muon corrections to be discussed later. Figure 2b shows the TOF of the elastic reaction candidates after these cuts. A clear prompt signal is observed above background.

High energy neutrons (momenta > 1 GeV/c) however, will also be in the RF buckets and cannot be separated by TOF. As part of our study of neutral current final states we have also studied the reaction $\nu_\mu p \to \nu_\mu p \pi^o$. This reaction is quite sensitive to $np \to np\pi^o$ contamination. Based on the observed distribution of the π^o events along five interaction lengths we conclude that the high energy neutron contamination is insignificant.

After imposing the cuts discussed above we obtain 39 events which satisfy the criteria for reaction (1). After imposing a flat background subtraction we obtain 25+6 events in our final sample. Using the same cuts for the quasi-elastic events as we have imposed on the neutral current candidates, we observe 65 quasi-elastic events.

We now consider background arising from the single pion reactions:

$\nu n \to \mu^- p \pi^o$ (3a) $\nu p \to \nu p \pi^o$ (4a)

$\nu p \to \mu^- p \pi^+$ (3b) $\nu n \to \nu p \pi^-$ (4b)

$\nu n \to \mu^- n \pi^+$ (3c) $\nu p \to \nu n \pi^+$ (4c)

Reactions 3(4) can be misidentified as reaction 2(1) if either the π or p is missing and the other track falls in the same range as the proton in reaction 2(1).

The misidentification of charged current single pion production events can be estimated from published cross sections[5]. It has been established that $\sigma(\nu n \to \mu^- p) \simeq \sigma(\nu p \to \mu^- p \pi^+)$. The relative ratio for reactions 3a, 3b, and 3c is 4:8:1 after correcting for the charge exchange effects in the complex nucleus[6]. Using our measured efficiencies for detection of a proton or pion in the final state we estimate that 7 events classified as $\nu n \to \mu^- p$ are due to pion contamination. This is consistent with the fact that ~10% of the events in Figure 1 are not coplanar.

In order to estimate the single pion events misidentified as elastic neutrino scattering we use a ratio independently measured in this experiment:

$$\frac{\sigma(\nu p \to \nu p \pi^o)}{\sigma(\nu n \to \mu^- p \pi^o)} = 0.16 \pm 0.04 \ .$$

We now consider the ratios:

$$\frac{\sigma(\nu n \to \nu p \pi^-)}{\sigma(\nu p \to \nu p \pi^o)}$$

and

$$\frac{\sigma(\nu p \to \nu n \pi^+)}{\sigma(\nu p \to \nu p \pi^o)}$$

If the I-spin of the πN system is 1/2, then the ratio of the above cross sections is 2. This is the worst possible case. After a correction for the detection efficiencies of the final state particles, we estimate that the background due to these reactions is

< 4 events.

The other major background source for reaction (1) comes from reaction (2) due to an inefficiency of the detector in detecting wide angle muons. In order to correct for this we have made a Monte Carlo calculation of reaction (2) using standard V-A theory[7]. By comparing the observed muon angular distribution with the calculated muon angular distribution, Figure 3, we find that 8±2% of the muon in the quasi-elastic reaction escapes detection in the detector. We estimate, therefore, that 5 such wide angle muon events have been misidentified as reaction (1).

The total background after neutron subtraction in the final 25 events identified as reaction (1) is less than 10 events. We conclude, therefore, that we have observed elastic neutrino proton scattering. With all the corrections discussed above, we obtain the ratio:

$$\frac{\sigma(\nu p \to \nu p)}{\sigma(\nu n \to \mu^- p)} = 0.25 \pm 0.12 \ .$$

The kinematic range for the above ratio is $0.3 < q^2 < 1.0 (GeV/c)^2$.

REFERENCES

1. The members of this collaboration are W. Lee, E. Maddry, P. Sokolsky, L. Teig, G. Alverson, A. Bross, T. Chapin, L. Holloway, L. Nodulman, T. O'Halloran, C. Pang, K. Goulianos and L. Litt.
2. F. J. Hasert et al., Phys. Lett. 46B, 138 (1973).
   A. Benvenuti et al., Phys. Rev. Lett. 32, 800 (1974).
   S. J. Barish et al., Phys. Rev. Lett. 33, 448 (1974).
   B. C. Barish et al., Phys. Rev. Lett. 34, 538 (1975).
   W. Lee et al., paper contributed to the XVII International Conf. on High Energy Physics, London, 1974.
3. D. C. Cundy et al., Phys. Lett. 31B, 478 (1970).
   S. J. Barish et al., paper contributed to the XVII International Conf. on High Energy Physics, London, 1974.
4. A. Salam, in Elementary Particle Theory, edited by N. Svartholm (Almquist and Forlag, Stockholm, 1968), p. 367; S. Weinberg, Phys. Rev. Lett. 19, 1264 (1967); Phys. Rev. D5, 1412 (1972).
5. P. Schreiner in Neutrinos-1974, AIP Conference Proceedings No. 22 edited by C. Baltay, p. 101.
6. S. L. Adler, S. Nussinov, and E. A. Paschos, Phys. Rev. D9, 2125 (1974).
7. See, for example, C. H. Llewellyn-Smith, Physics Report 3C, 261 (1972).
8. For a more detailed reference, see J. J. Sakurai and L. F. Urrutia, Phys. Rev. D11, 169 (1975).

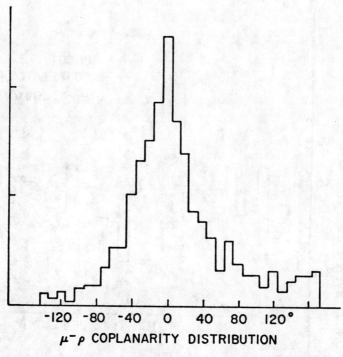

$\mu^- \rho$ COPLANARITY DISTRIBUTION

Figure 1

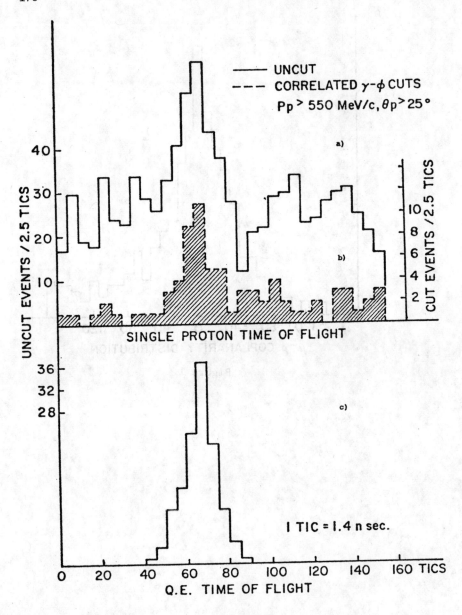

Figure 2

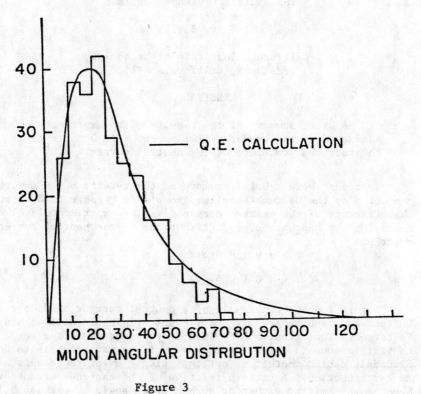

Figure 3

THE NEUTRAL CURRENT COUPLINGS*

B. C. Barish

California Institute of Technology
Pasadena, California 91125

ABSTRACT

A brief summary of the results of an experiment conducted by the Caltech-Fermilab group in October 1974 to study the structure of the neutral current.

I present here a brief summary of the results of an experiment conducted by the Caltech-Fermilab group[1] in October 1974 to study the structure of the neutral current.[2] The experiment involved measuring the hadron energy distributions in the neutral current reactions

$$\nu + N \to \nu + \text{hadrons}$$

and
$$\bar{\nu} + N \to \bar{\nu} + \text{hadrons}.$$

Figure 1 shows an example of a neutral current candidate in the Caltech-Fermilab apparatus. A ν-event originates well within the Fe target, significant energy in a hadron cascade is recorded, but no visible muon is observed. However, it is not possible to unambiguously decide whether or not this single event represents a neutral current reaction. A certain fraction of the charged current events have a muon emitted either at such a large angle or with such low energy that the muon is not observed.

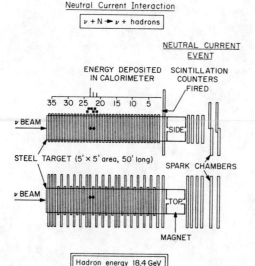

Fig. 1: A neutral current candidate in the Caltech-Fermilab apparatus is shown. A neutrino enters from the left, interacts well within the iron target, showing a significant deposit of energy into hadrons. There are no indications of a muon in the final state.

For the experiment reported here the charged current distributions have been studied in detail using the events where the muon was clearly identified. In order to analyze the neutral current data the charged current analysis was used for two main purposes:

(1) Determination and subtraction of the charged current background in the sample of neutral current candidates.

(2) Determination of the relative normalization of neutrino and antineutrino data. Direct <u>external</u> normalization by monitoring the beam by procedures used in <u>our</u> normalized cross section work(3) was not possible for this measurement because of the short spill running (~ 1 msec).

Figure 2 shows the charged current energy distributions for events where the muon traversed the magnet. The two peak structure of the dichromatic beam is apparent for the neutrino data. However, for antineutrinos the relatively low K⁻ flux is reflected in the smaller number of high energy antineutrino events.

The charged current data was analyzed and fit to three <u>radically</u> different hypothesis in order to determine the sensitivity of the neutral current conclusions to the physics of charged currents.

Fig. 2

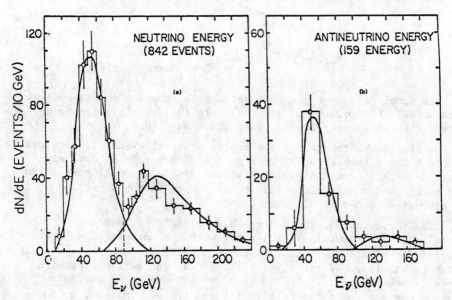

CITF DATA- DISTRIBUTIONS IN TOTAL OBSERVED
ENERGY NARROW BAND BEAM CHARGED-CURRENTS

Model I: Scaling Model -

$$\frac{d\sigma^{\nu}}{dy} = \frac{G^2 ME_{\nu}}{\pi} \left[q(x) + \bar{q}(x)(1-y)^2 \right]$$

$$\frac{d\sigma^{\bar{\nu}}}{dy} = \frac{G^2 ME_{\bar{\nu}}}{\pi} \left[\bar{q}(x) + q(x)(1-y)^2 \right]$$

$$q(x) + \bar{q}(x) = F_2^{ed}(x)$$

$$\bar{q}(x) = \frac{F_2^{ed}}{2}(x) \, e^{-\lambda x}$$

In this model we parameterize <u>all</u> our ignorance about charged currents in one parameter - $\alpha = \bar{Q}/(Q+\bar{Q})$ (where $Q = \int_0^1 q(x)\,dx$, etc.).

Model II: Non-scaling model - In this case we allow non-scaling energy dependence by letting $\alpha \to \alpha(E_{\nu})$. This means we make the same assumptions as in Model I except that now there are two ignorance parameters α_1 and α_2 (one for pion decay neutrinos, $E_{\nu} \sim 50$ GeV, and one for kaon decay neutrinos, $E_{\nu} \sim 150$ GeV).

Model III: Non-scaling model with right handed currents and production of a new heavy quark.(4)

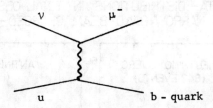

In this model we assume the best value of α from low energy neutrino data and electroproduction ($\alpha \sim 0.06$). In this case the ignorance parameter (free parameter) is the heavy quark <u>mass</u> (M_b).

A minimum hadron energy for neutral current events is required by the trigger (see Fig. 3). For good efficiency, only data with $E_H > 12$ GeV has been analyzed. Several additional small corrections and background subtractions have been applied to the data, and will be discussed in detail elsewhere.

The corrected neutral current hadron distributions for $E_H > 12$ GeV are shown in Fig. 4 along with the corresponding charged current distributions. The low energy cut-off plus our lack of detailed knowledge of the physics of the charged currents at high energies make the ratios R_{ν} and $R_{\bar{\nu}}$ of observed neutral current to charged current events not easily interpretable. Instead, we have

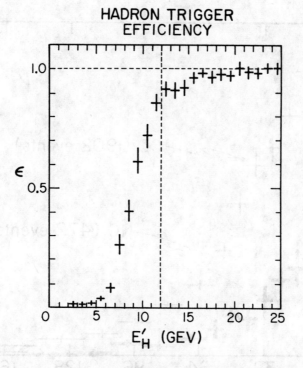

Fig. 3

The detection efficiency for observing a hadron shower in the apparatus vs. the energy of the hadrons is shown.

directly analyzed our neutral current data. The charged current were used only to determine the ratio of neutrino to antineutrino fluxes (for normalization) and to determine the charged current subtraction from the neutral current sample.

The neutral currents have then been analyzed using the expression

$$\frac{d\sigma^\nu}{dy} = \left(\frac{G^2 ME}{\pi}\nu\right)\left(\int F_2^{ed}(x)dx\right)\left[g_N + g_P(1-y)^2 + g_F y^2\right]$$

$$\frac{d\sigma^{\bar\nu}}{dy} = \left(\frac{G^2 ME}{\pi}\bar\nu\right)\left(\int F_2^{ed}(x)dx\right)\left[g_P + g_N(1-y)^2 + g_F y^2\right]$$

g_N, g_P, and g_F represent the coefficient of negative helicity, positive helicity and helicity flip. For this report I will only discuss the data under the assumptions that $g_F \equiv 0$ and that the neutral currents scale and rise linearly with energy. Both these assumptions are still being investigated.

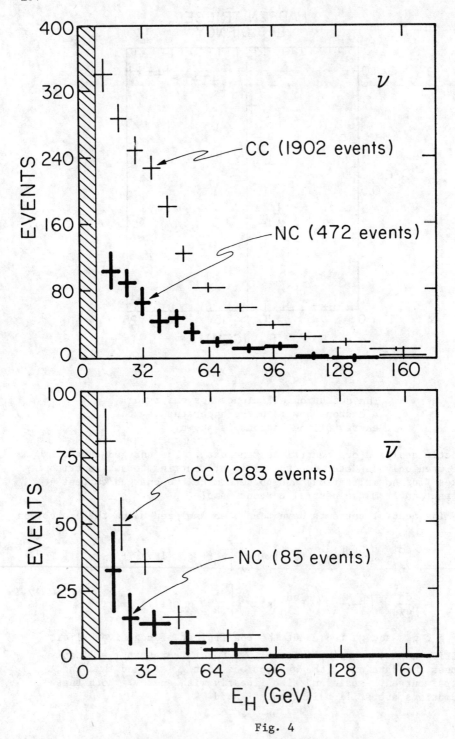

Fig. 4

Charged and neutral current distributions for neutrinos and antineutrinos with $E_{had} > 12$ GeV.

We have made a two parameter fit for g_N and g_P or alternately

$$g = g_N + g_P \quad \text{and} \quad P = g_P/g$$

The results using the three different fits to the charged current data are given below:

	CC Model	Ignorance Parameter	χ^2	g	P
I	Scaling	$\alpha = 0.27 {}^{+.08}_{-.13}$	18.6	$.30 \pm .03$	$.36 \pm .09$
II	Non-Scaling	$\alpha_1 = 0.17 {}^{+.13}_{-.11}$ $\alpha_2 = .32 {}^{+.18}_{-.15}$	15.8	$.30 \pm .03$	$.39 \pm .09$
III	B - Quark $\alpha = .06$	$M_B = 5 \pm 1.5$ GeV	20	$.32 \pm .04$	$.33 \pm .09$

Note that the answers for g and P are relatively insensitive to the physics of the charged currents. Figure 5 shows the contours for our best two parameter fit for g_N and g_P. Including systematic errors these best fits correspond to

$$g = (.30 \pm .04) \pm .02 \quad \text{or} \quad g_P = (.11 \pm .04) \pm .02$$
$$P = (.36 \pm .09) \pm .04 \qquad \qquad g_N = (.19 \pm .02) \pm .02$$

statistical ↙   ↘ estimated
error   systematic
              error

These results give the strength and fraction of positive helicity in the neutral current sample. It must be note, however, that the positive coefficient represents a combination of any V+A interaction and scattering of neutrinos off antiquarks. In terms of the V-A (g_-) and V+A (g_+) coefficients we can re-express g_N and g_P as follows:

$$g_N = (1-\alpha) g_- + \alpha g_+$$
$$g_P = \alpha g_- + (1-\alpha) g_+$$

The determination of g_- and g_+ requires a knowledge of α (fraction of momentum carried by antiquarks in the nucleon). Figure 6a and 6b show the two parameter contours for g_- and g_+ assuming $\alpha = 0.06$ (best fit at low energies) and $\alpha = 0.17$ (best fit at ~50 GeV).

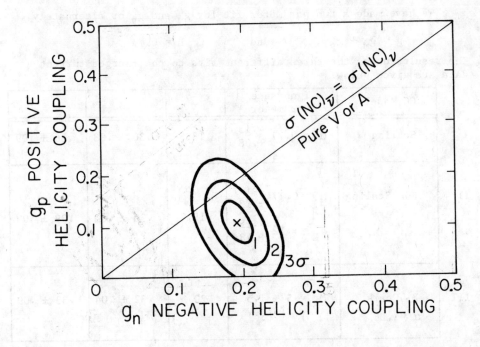

Fig. 5

Two parameter contours for the best fit for the negative and positive helicity coupling for neutral currents.

From these distributions g_- and g_+ we see the conclusions are reasonably insensitive to the fraction of antiquarks. As is shown, the results are quite consistent with the Weinberg-Salam model[5] both in magnitude and in V,A mixing parameter for $\sin^2 \theta_\omega \sim 0.3 - 0.4$ (the best value is still being determined). Our best fit lies between V - A and V, however it is within about 1.5 standard deviations of pure V (or A) and also is about 1 - 1.7 standard deviations from V - A (depending on the amount of $\bar{q}$ in the nucleon).

Overall, we are very encouraged by this first attempt to make quantitative measurements of the strength and type of coupling for the neutral current interaction. Information on possible S, P, or T contributions or non-scaling behavior should be forthcoming. Also, with improvements now underway, much more precise measurements will be possible in the future.

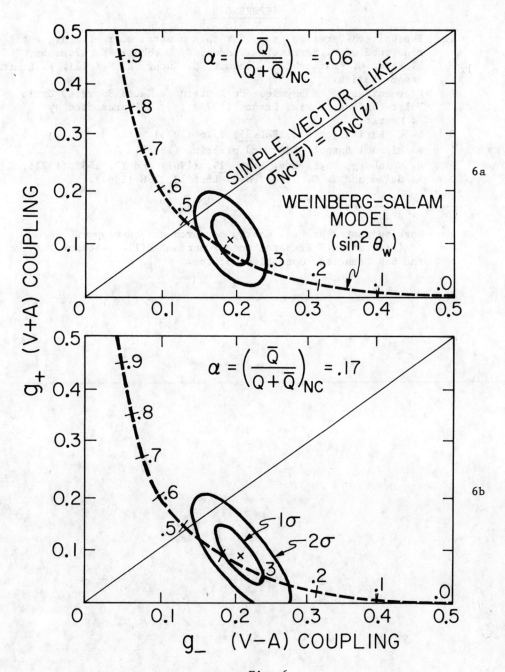

Fig. 6

Two parameter contours V−A vs. V+A are shown for α = .06 (6a) and α = .17 (6b). Predictions of the Weinberg-Salam model and simple vector-like theories are shown for comparison.

REFERENCES

1. The Caltech-Fermilab group for this neutral current measurement consisted of F. Merritt, A. Bodek, D. Buchholz, K. W. Brown, E. Fisk, G. Krafczyk, F. Jacquet, H. Suter, F. Sciulli, L. Stutte, and B. Barish.
2. Proceedings of La Physique du Neutrino a Haute Energie, Ecole Polytechnique, Paris, France (1975) (pg. 291 presented by F. Merritt).
3. B. C. Barish, et al., PRL $\underline{35}$, 1316 (1975).
4. R. Michael Barnett, Harvard preprint (1975).
5. S. Weinberg, Phys. Rev. Lett. $\underline{19}$, (1967) and $\underline{27}$, 1683 (1972), A. Salam and J. C. Ward, Phys. Lett. $\underline{13}$, 168 (1964).

---

* Work supported by U.S. Energy Research and Development Administration. Prepared under Contract E(11-1)-68 for the San Francisco Operations Office.

ANOMALIES IN $\nu, \bar{\nu}$ INTERACTIONS

A. Benvenuti, D. Cline, W. T. Ford, R. Imlay,*
T. Y. Ling, A. K. Mann, R. Orr,+ D. D. Reeder, C. Rubbia,++
R. Stefanski, L. Sulak, and P. Wanderer
University of Wisconsin, Madison, Wis. 53706

*Now at Rutgers University, New Brunswick, N. J. 08903

+Now at CERN, Geneva 23, Suisse

++Now at Brookhaven National Laboratory, Upton, N. Y. 11973

Presented by A. Benvenuti

ABSTRACT

The state of the large y anomaly is evaluated following a detailed study of the resolution and the detection efficiency of the apparatus. A new measurement of the cross-section ratio $\sigma_c(\bar{\nu})/\sigma_c(\nu)$ at high energies is presented.

INTRODUCTION

In the past year there have been many reports of new phenomena in neutrino and antineutrino interactions. The observation of events with two muons in the final state:

$$\nu + N \rightarrow \mu^-\mu^+ X (\nu)$$

$$\bar{\nu} + N \rightarrow \mu^+\mu^- X (\bar{\nu})$$

first reported by the HPWF collaboration[1] has since been confirmed by the CITF group.[2] More recently two bubble chamber experiments[3,4] have observed the following reactions:

$$\nu + N \rightarrow \mu^- e^+ X (\nu)$$

$$\bar{\nu} + N \rightarrow \mu^+ e^- X (\bar{\nu})$$

which are naturally associated with the dimuon events by µe universality.

We have also presented data on the scaling variable distributions[5] which showed considerable deviations from the scaling predictions for the antineutrino events at large energies (large y anomaly).

It is very tempting to make a connection between dimuon events and the µe events with the large y anomaly. In particular if the extra lepton is the product of the semileptonic decay of a new particle we would also expect deviations in the scaling variable distributions from the hadronic decay modes of this new particle.

The large y anomaly has not yet been confirmed or disproved by other experiments and it has been the object of some controversy. Therefore we will address here the question of the experimental validity of the large y anomaly and its characteristics. We will also present a new determination of the $\sigma_c(\bar{\nu})/\sigma_c(\nu)$ ratio at high energies.

EXPERIMENTAL DETAILS

1. Description of the apparatus.

The HPWF apparatus, shown in Fig. 1, consists of a liquid scintillator calorimeter, hadron filter, muon spectrometer and trigger counters (A, B, C, D).

The calorimeter is subdivided into 16 modules 10' x 10' x 18" each, thus allowing for the determination of the interaction origin along the beam line within 9". Optical spark chambers are inserted every four modules to visualize the hadronic shower. The calorimeter performs the following functions:
  a. Target: using a restricted fiducial volume of 20 metric tons from module 2 to module 10.
  b. Detector: the full 60 tons are used to measure the energy deposited by the hadronic shower.
  c. Trigger: interactions producing hadronic energy in excess of a prefixed threshold, usually set at 2 GeV, are taken as good triggers.

The long term stability of the calorimeter was monitored continuously by recording the response to high energy muons associated with the neutrino beam.

The muon spectrometer consists of four iron toroids 6' in radius and 4' thick. Optical spark chambers are placed in front and after each toroid to visualize the muon track.

The events were selected using two complementary triggers: $\overline{A}BCD$, $\overline{A}E$. The first trigger requires a muon in the final state to traverse the entire spectrometer and thus puts a severe restriction on the angular acceptance of the apparatus. The second trigger requires only energy deposition in the calorimeter and thus does not require any muon in the final state. The combined trigger is biased only at very low energy, therefore events with energy less than 10 GeV were not included in the present analysis.

2. Resolution

We want to study the effects of resolution on the dimensionless variables $y = E_H/E_\nu$ and $x = Q^2/(2mE_H)$ with $Q^2 = 4E_\nu E_H \sin^2(\theta_\mu/2)$ and $E_\nu = E_H + E_\mu$. The errors on these variables are related to the errors on the measured quantities E_H, P_μ, θ_μ in the following way:

$$\left(\frac{\delta x}{x}\right)^2 = 4\left(\frac{\delta\theta_\mu}{\theta_\mu}\right)^2 + (2-y)^2 \left(\frac{\delta P_\mu}{P_\mu}\right)^2 + (1-y)^2 \left(\frac{\delta E_H}{E_H}\right)$$

$$\left(\frac{\delta y}{y}\right)^2 = (1-y)^2 \left[\left(\frac{\delta E_H}{E_H}\right)^2 + \left(\frac{\delta P_\mu}{P_\mu}\right)^2\right]$$

From these equations it is apparent that the y variable is measured with greater accuracy than the x variable. Furthermore the resolution in y at large y improves because the $(1-y)^2$ factor vanishes.

The resolution in the measured variable was determined by measuring the response of the apparatus to pion and muon beams with known energy between 17 and 150 GeV. The results of this calibration have been published elsewhere.[6] Typically $\delta E_H/E_H$ is approximately constant and of the order of 12% above 20 GeV, $\delta P_\mu/P_\mu$ varies from 15% to 20% in the energy range considered and $\delta\theta_\mu \sim 5$ mrad at $P_\mu \sim 20$ GeV/c. The resulting (x,y) resolution is shown in Fig. 2.

We have investigated the effects of resolution smearing in the (x,y) plane by using a trial function for the (x,y) dependence of the events, the experimentally determined resolution in the measured variables and Monte Carlo techniques. The resulting smearing proves to be quite severe in the x variable but not in y. This is illustrated by considering the effects of resolution smearing in a single (x,y) bin, Fig. 3. Typically the value of y is not changed by more than one bin (±.1), while x can be pushed to unphysical values.

The effects of resolution are particularly important when the data are presented after cuts are made in x. To illustrate this point we show in Fig. 4a the y distribution as smeared by resolution together with the input distribution used for $0 \leq x < 0.15$ and $10 \leq E < 30$ GeV. The depletion of events at small y is caused by the cut in x since in the small x range considered the resolution shifts events preferentially to larger x values. This effect is considerably less pronounced if a larger x range is used (Fig. 4c) or higher energies are considered (Fig. 4b,d).

In conclusion we have found no indication that the experimental resolution can create an artificial excess of events at large y.

3. Acceptance

The acceptance of the apparatus is determined by two factors: the minimum muon momentum measurable and the maximum muon angle observable in the apparatus. These in turn are dictated by the fact that a muon must traverse at least one magnet section in order to be measurable. In the HPWF apparatus the maximum muon angle for an interaction occurring in the middle of the fiducial volume is 225 mrad and the minimum muon momentum is 5 GeV/c if we disregard the effects of the magnetic focusing. The maximum value of y observable in function of x can be calculated for each neutrino energy as:

$$(1) \begin{cases} y_{max} = k/(x + k) \\ k = 2E_\nu \sin^2(\theta_\mu/2)/m \end{cases}$$

The resultant contour for two incident neutrino energies are shown in Fig. 5. In this schematization events with (x,y) values above the contours (blind region) are not observable. The size of the blind region is energy dependent and decreases with increasing neutrino energies. In the actual experimental situation the contours will be smeared into bands by the random occurrence of the events in the fiducial volume and by the focusing properties of the magnet.

THE DATA

The data presented here consists of about 5000 ν and 2500 $\bar{\nu}$ events with a single muon in the final state and were accumulated using very different beams and under different magnet focusing configurations. The resulting energy distributions are shown in Fig. 6a,b for ν and $\bar{\nu}$ respectively.

In the present analysis each event has been corrected for the geometrical acceptance of the apparatus in a model independent way by rotating the muon track in the magnet and weighting the event according to the fraction of the azimuthal angle visible. Events with weight greater than 5 were not included in the sample. We also have not attempted to extract the experimental resolution from the data, rather we have smeared the theory according to the observed resolution whenever necessary.

THE LARGE y ANOMALY

In the scaling region the differential cross-section for inclusive scattering is given by

$$(2) \quad \frac{d\sigma}{dx})^{\nu,\bar{\nu}} = \frac{G^2 ME}{\pi} \int F_2^{\nu,\bar{\nu}}(x) dx [1-y(1-\frac{y}{2})(1\pm B^{\nu,\bar{\nu}}) + \frac{y^2}{2} R_L^{\nu,\bar{\nu}}]$$

with
$$B^{\nu,\bar{\nu}} = -\int xF_3^{\nu,\bar{\nu}}(x)dx / \int F_2^{\nu,\bar{\nu}}(x)dx$$

$$R_L^{\nu,\bar{\nu}} = [\int (2xF_1^{\nu,\bar{\nu}}(x) - F_2^{\nu,\bar{\nu}}(x)dx) / \int F_2^{\nu,\bar{\nu}}(x)dx]$$

For interactions occurring on isoscalar targets charge symmetry invariance requires that $F_i^{\nu}(x) = F_i^{\bar{\nu}}(x)$. In the case of pure V-A interactions on spin 1/2 constituents and no antiquark components in the nucleon we have $B = 1$ and $R_L = 0$ and equation 2) reduces to:

$$3) \quad \frac{d\sigma}{dy})^{\nu} = \frac{G^2 ME}{\pi} \int F_2(x) dx$$

$$4) \quad \frac{d\sigma}{dy})^{\bar{\nu}} = \frac{G^2 ME}{\pi} \int F_2(x) dx \, (1-y)^2$$

The y distributions in the energy region $10 \leq E < 30$ GeV are shown in Fig. 7a,b respectively. These distribution as the followings are given for x values less than 0.6 in order to insure both good efficiency and resolution. Equation 2) with $R_L = 0$ was used to fit the data in the region $0.1 \leq y \leq 0.55$ and gave the values of $B^{\nu} = 0.9 \pm .6$ and $B^{\bar{\nu}} = 0.95 \pm .1$. The cut off in y was chosen to eliminate the effects of the blind region on the determination of B. Fig. 8a,b show the corresponding distribution for E > 50GeV. The neutrino data are fit by a value of $B^\nu = 0.75 \pm .2$ in the region of full efficiency and good resolution ($0.1 \leq y \leq 0.85$). In contrast the anti neutrino data are fit by a value of $B^{\bar\nu} = 0.45 \pm 0.15$ indicating a large departure from the $(1-y)^2$ dependence expected from scaling as stated above.

In the quark parton model the B parameter is simply related to the antiquark content of the nucleon by

5) $\quad B = \int [q(x) - \bar{q}(x)]dx / \int [q(x) + \bar{q}(x)]dx$

6) $\quad \alpha = \int \bar{q}(x) \, dx / \int [q(x) + \bar{q}(x)] \, dx = (1-B)/2$

From the above values of B we obtain for the antineutrinos the value of $\alpha^{\bar\nu} = 0.03 \pm .05$ in the energy region 10<E<30 GeV in good agreement with the value of 0.1 reported by other experiments.[7,8] However for E>50 GeV the fit yields a value of $\alpha^{\bar\nu} = 0.3 \pm 0.08$ which is too large to be explained by the simple quark-parton model and in contradiction with the measurement at lower energy. We are forced therefore to conclude that the large y anomaly cannot be explained in the conventional models.

To study the energy dependence of the effect in greater detail we show in Fig. 9a,b the average value of y, <y>, for ν and $\bar\nu$ respectively together with the Monte Carlo predictions for B = 0.9 and 0.4 which take into account the finite acceptance of the apparatus and the experimental resolution. The ν data are not sensitive to the particular value of B chosen and can be fit in the whole energy region by either B value. In contrast the $\bar\nu$ data show a sharp rise of <y> with energy from a value corresponding to B = 0.9 below E $\sim$ 30 GeV to a plateau value corresponding to B = 0.4 above E $\sim$ 60 GeV. Again the data cannot be fit by a unique value of B over the energy range considered indicating that although the B parameterization can be used as an indication of the large y anomaly it is not able to describe the phenomenon in a satisfactory way.

We have also studied the dependence of the large y anomaly on x by dividing the data of Fig. 8b into two x regions: 0.--0.15 and 0.15--0.6. The resulting distributions are shown in Fig. 10a and b respectively. The anomaly appears to be concentrated in the low x region which yields a value of B = 0.2 ± 0.3 while in the high x region the y distribution is very closely represented by 4) and gives a value of B = 0.8 ± 0.1. The xy correlation of the large y anomaly is shown in more detail in Fig. 11a where the y distributions for $\bar{\nu}$ are given for E > 30 GeV in x bins of 0.1 width. Again the anomaly appears to be concentrated in the lowest x bins. Conversely the x distribution for $\bar{\nu}$ in the same energy region appear to become sharper at large y values, Fig. 11b.

It is interesting to note that the dimuon events[1] appeared also to be produced preferentially at small x and large y suggesting that the large y anomaly and the dimuon events are different manifestations of the same phenomenon.

In conclusion the large y anomaly in the $\bar{\nu}$ data has been verified using a sample 10 times larger than in our earlier report.[5] The ν data do not show evidence for or against the presence of a large y anomaly. The characteristics of the anomaly suggest that new particle(s) are produced with an apparent threshold of $\sim$30 GeV and support the hypothesis that both the dimuon events and the large y anomaly come from a common source.

DETERMINATION OF $\sigma_c(\bar{\nu})/\sigma_c(\nu)$ AT HIGH ENERGIES

The worst limitation in the determination of the $\sigma_c(\bar{\nu})/\sigma_c(\nu)$ rests with the incoming ν and $\bar{\nu}$ fluxes which are in general poorly known. This limitation is most severe when pulsed sign-selecting devices such as the magnetic horn are used to focus the secondary particles which generate the $\nu,\bar{\nu}$ beams. In these beams the $\bar{\nu}$ to ν flux ratio depends critically on the knowledge of the focussing devices and cannot be calculated reliably by Monte Carlo techniques.

However beams like the Quadrupole Triplet[9] which do not apply a sign selection to the secondaries and are not pulsed can be calculated quite reliably. The only limitation in this case stems from the fact that the $\bar{\nu}$ events are considerably less than the ν events since the fluxes are not enhanced.

We want to report here two determinations of the $\sigma_c(\bar{\nu})/\sigma_c(\nu)$ ratio based on two independent methods.

1. Flux Dependent Method

In this analysis we have used only a sample of 570 $\bar{\nu}$ and 2900 ν events taken with a primary proton energy of 380 GeV and a quadrupole-triplet beam set to focus 200 GeV secondaries. The $\nu,\bar{\nu}$ fluxes for this beam are shown in Fig. 12 and were calculated using the measured yields of hadrons[10] and the focussing properties of the quadrupole triplet beam. The fluxes up to 80 GeV are dominated by pion neutrinos and anti-neutrinos, this is an important factor in determining the cross section ratio since the π^+/π^- ratio is better known that the K^+/K^- ratio. The $\bar{\nu},\nu$ flux ratio is shown in Fig. 13 together with the corresponding ratio for an unfocussed beam. Above $\sim$ 30 GeV the flux ratios are not different for the two beams indicating that the $\bar{\nu}/\nu$ flux ratio is not critically dependent on the knowledge of the focussing properties of the quadrupole triplet beam as anticipated.

The $\sigma_c(\bar{\nu})/\sigma_c(\nu)$ ratio was determined using the $\nu,\bar{\nu}$ fluxes and the corrected number of ν and $\bar{\nu}$ events. In this case the observed events were corrected not only for the geometrical efficiency but also for the loss of events due to the finite acceptance of the apparatus in a model dependent way. The size of this correction is rather small at energies above 50 GeV as can be seen in Table 1.

The energy dependence of the $\sigma_c(\bar{\nu})/\sigma_c(\nu)$ ratio is shown in Fig. 14. At energies below 50 GeV the ratio is in good agreement with previous measurements[11,12] whereas at larger energies the cross section ratio rises above 0.4.

2. Flux Independent Method.

This method is based on the Lee-Yang theorem[13]

$$\lim_{E \to \infty} \frac{d\sigma}{dW^2} (\bar{\nu} + N \to \mu^+ + X) =$$

$$= \lim_{E \to \infty} \frac{d\sigma}{dW^2} (\nu + N \to \mu^- + X)$$

which depends only on charge symmetry invariance and holds for isoscalar targets. At finite energies the above theorem is valid as long as the hadronic mass W of X is less than a prefixed value W_{max}. Sakurai[14] has indicated the range of W values in function of energy for which the theorem is satisfied to better than 15%.

This prescription can be used to determine the $\bar{\nu}/\nu$ flux ratio from the data themselves independently from the knowledge of the beam characteristics.

The $\sigma_c(\bar{\nu})/\sigma_c(\nu)$ ratio obtained using the full sample of the data and the above technique for the $\bar{\nu}/\nu$ flux determination is shown in Fig. 15 for several choices of W_{max}. The resulting $\sigma_c(\bar{\nu})/\sigma_c(\nu)$ does not depend critically on the particular W_{max} used and it is in good agreement with the values obtained in the flux dependent method.

In conclusion we have found that the ratio of the cross sections grows with energy reaching the value of 0.6 below 100 GeV. The growing value of the cross section ratio with energy is probably related with the large y anomaly and the dimuon events.

REFERENCES

1) A. Benvenuti, et al., Phys. Rev. Letters 34, 419, 597 (1975), Phys. Rev. Letters 35, 1199 (1975), Phys Rev. Letters 35, 1203 (1975), Phys. Rev. Letters 35, 1249 (1975).

2) B. Barish, et al., Proceedings of the Paris International Conference on Neutrino Interactions, Ecole Polytechnic, p. 131.
B. Barish, Int'l Conference on Lepton and Photon Interactions at High Energy, Stanford University, Aug. 1975.

3) J. von Krogh, et al., "Observation of $\mu^- e^+ K_s^0$ Events Produced by a Neutrino Beam," Wisconsin preprint.

4) F. Nezrick, A.P.S. Meeting, New York, 1976.

5) A. Aubert, et al., Phys. Rev. Letters 34, 984 (1974).

6) A. Benvenuti, et al., Nucl. Instr. Method 125, 447 (1975).
A. Benvenuti, et al., Nucl. Instr. Method 125, 457 (1975).

7) H. Deden, et al., Nucl. Phys. B85, 269 (1975).

8) L. Mo, Int'l Conference on Lepton and Photon Interactions at High Energy, Stanford University, Aug. 1975.

9) A Description of the Fermilab quadrupole triplet neutrino beam is given in A. Benvenuti, et al., Proceedings of the Paris International Conference on Neutrino Interactions, Ecole Polytechnic, p. 397 (1975).

10) B. Aubert, et al., Proceedings of the Paris International Conference on Neutrino Interactions, Ecole Polytechnic, P. 385 (1975).
W. L. Baker, et al., Phys. Letters B51, 303 (1974).

11) T. Eichten, et al., Phys. Rev. Letters B46, 274 (1973).
A. Benvenuti, et al., Phys. Rev. Letters 30, 1084 (1973).
A. Benvenuti, et al., Phys. Rev. Letters 32, 125 (1974). The ratio of $\sigma_c^{\bar{\nu}}/\sigma_c^{\nu}$ was measured to be 0.34 ± 0.03 at an average energy of 27 GeV. B. C. Barish, et al., Phys. Rev. Letters 35, 1316 (1975).

12) R. Imlay, et al., London Conference (1974). The $\sigma^{\bar{\nu}}/\sigma^{\nu}$ ratio of 0.4 ± 0.1 was measured using quasi-elastic flux normalization. The average neutrino energy was 30 GeV.

13) T. D. Lee and C. N. Yang, Phys. Rev. 126, 2239 (1962).

14) J. J. Sakurai, Flux Determination in Neutrino Experiments at NAL Energies, UCLA preprint UCLA/73/TEP/79 (1973) (unpublished).

Table I

E_ν (GeV)	ε_ν	$\varepsilon_{\bar{\nu}}$
10-30	0.56	0.88
30-50	0.75	0.96
50-70	0.82	0.98
70-90	0.86	0.99

Table I: Correction for events that fall outside the acceptance of the detector, $N_{corr.}^{\nu,\bar{\nu}} = N_{obs.}^{\nu,\bar{\nu}}/\varepsilon_{\nu,\bar{\nu}}$.

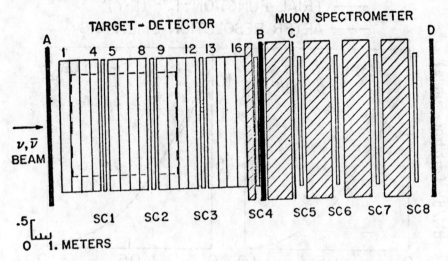

Fig. 1

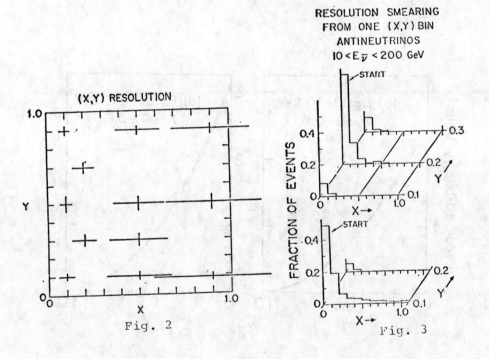

Fig. 2

Fig. 3

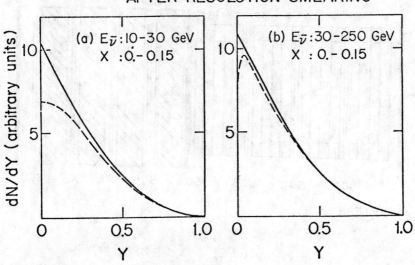
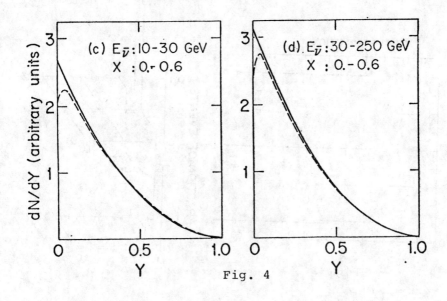

Fig. 4

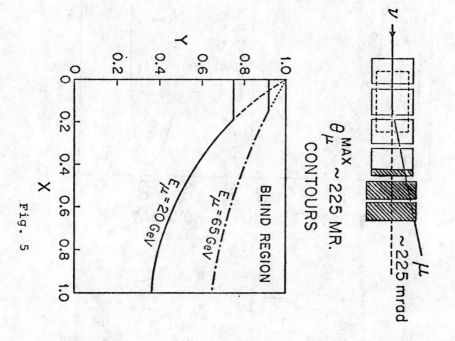

Fig. 5

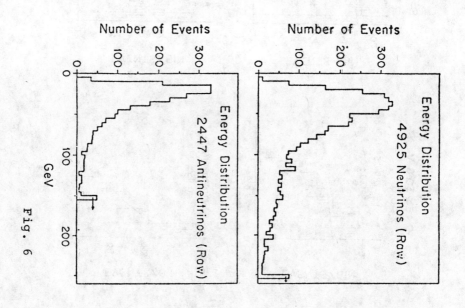

Fig. 6

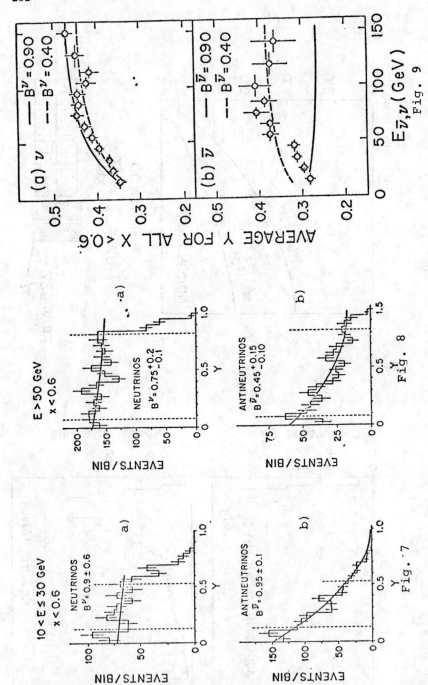

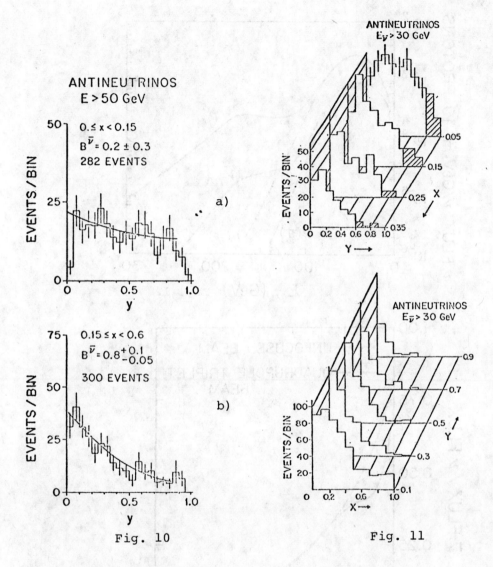

Fig. 10

Fig. 11

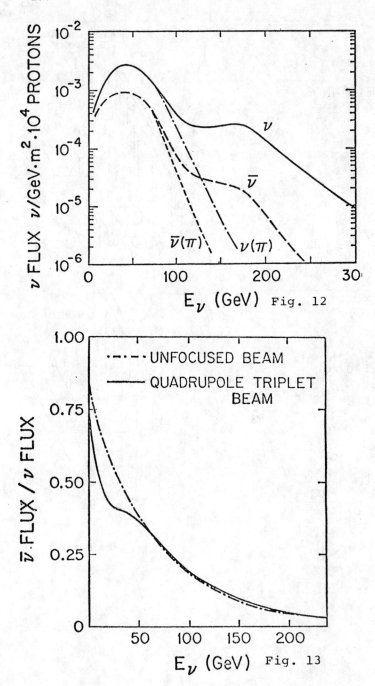

Fig. 12

Fig. 13

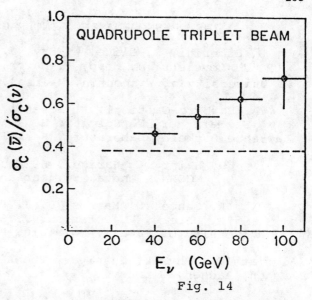

Fig. 14

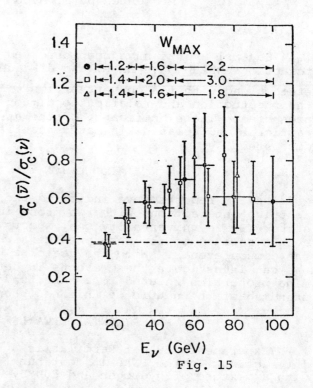

Fig. 15

NEW PHENOMENA IN ν INTERACTIONS

U. Camerini, D. Cline, W. Fry, J. von Krogh,
R. J. Loveless, J. Mapp, R. March, D. D. Reeder
University of Wisconsin, Madison, Wis. 53706

A. Barbaro-Galtieri, P. Bosetti,[+] G. Lynch,
J. Marriner, F. Solmitz, M. K. Stevenson
Lawrence Berkeley Laboratory, Berkeley, Cal. 94720

D. Haidt, G. Harigel, H. Wachsmuth
CERN, Geneva, Switzerland

R. Cence, F. Harris, S. I. Parker,
M. Peters, V. Peterson, V. Stenger
University of Hawaii, Honolulu, Hawaii 96822

[+]Also at III. Physik. Institut, Technische Hochschule Aachen, Aachen, Germany.

Presented by J. Mapp

ABSTRACT

Several events have been observed in neutrino interactions which have an e^+ and a K_s^o decay in the final state along with the conventional μ^-. A discussion of their detection and relation to dimuon events will be presented. The significance of the associated strange particles and related kinematic quantities will be discussed.

INTRODUCTION

Recent reports have indicated the observation of neutrino interactions with two muons in the final state.[1,2] By μe universality one expects to observe $\mu^- e^+$ events related to the same phenomena which produced the dimuon events. With the use of a bubble chamber one can investigate these events in detail to determine properties related to a more fundamental understanding of the nature of this new phenomena.

EXPERIMENTAL DETAILS

Experiment E28A at Fermilab is a study of neutrino interactions in the FNAL 15 foot bubble chamber filled with a mixture of hydrogen and 20% by volume of Neon.

The collaboration consists of the groups (Berkeley and Hawaii) who built the External Muon Identifier (EMI) and the Wisconsin and CERN groups. The bubble chamber can be thought of as a 12 foot diameter sphere with a 3 foot nose. The chamber is surrounded by the copper coils of a superconducting magnet which produce a 30 kilogauss magnetic field. The neutrino beam was produced by a 300 GeV/c proton beam of 6×10^{12} protons per pulse incident on a 12 inch Aluminum target with 2 horns set to focus positives. The neutrino flux drops by an order of magnitude from a maximum at 20 GeV to 60 GeV. The EMI consists of 24 1 meter x 1 meter wire planes set in 3 belts outside the main vacuum chamber. The vacuum space is filled with the coppper of the magnet coils and the remaining space with Zinc such that forward muons in the main fiducial volume pass through approximately 1.5 feet of Cu and Zn. This is 3 absorption lengths so muons with greater than 3 GeV/c momentum will penetrate and trigger the wire planes while only about 2% of the hadrons will "punch through". A muon is identified as a track from a neutrino interaction vertex which traversed the visible volume of the chamber without a kink or scatter and produced a "hit" or trigger in an EMI chamber within about 1 cm of the position predicted by extrapolating the track in question out through the Zinc and Copper coils to the space which the EMI plane occupy.

For neutrino physics this experiment has several unique features over previously reported experiments.

1) The bubble chamber allows a careful visual examination of the interaction vertex. An enlargement of the vertex enables one to look for close in decays, interactions, comptons, gamma ray conversions and asymetric dalitz pairs. In a bubble chamber strangeness can be detected by the observation of neutral strange particles through Vee decays such as K_S^o, K_L^o, Λ^o, and charged strange particles $K^{\pm}$ through their decays or interactions leading to the observation of a track with a kink with a Vee pointing to the kink.

2) The EMI allows the unique identification of leaving tracks as muons. It is then in principle easy to obtain a sample of events with 0 muons (neutral current) and 2 muons (dimuons) by looking for events with 0 or 2 EMI hits among the leaving tracks.

3) The H_2-Ne mixture provides many unique advantages.

First, the additon of a moderately heavy nucleous increases the mass of the target by a factor of 5 while the additonal neutrons increase the effective cross-section by an additional factor of 1.5 over that of a pure hydrogen run. For E28 there is a neutrino like interaction about every 8 pictures and an event with $E_{vis} > 10$ GeV every 10 to 12 pictures. The increased number of interactions makes it possible to look for rare processes at the level of a few percent and still have enough events to do a reasonable analysis.

Second, the increased cross-section for strong interactions makes it possible to separate hadrons from muons by looking for a scatter or interaction in a secondary track. This augments the EMI in the selection of muon tracks. The mean free path geometrical is approximately 2 meters.

Third, the addition of a heavy nucleous increases the detection of neutral energy. Neutrons and K_L^o will interact in the chamber. The conversion of gamma rays leads to the detection of π^o. For our mixture the radiation length is approximately 1 meter or 4 L_R across the chamber.

Fourth, the strong interaction of charged secondaries can aid in the identification of charged strange particles. Strangeness can not be lost so oftain a Vee will point to the secondary interaction vertex.

Fifth, the many radiation lengths in the chamber aid in the detection of directly produced e^+,e^- while the larger number of orbital electrons increase the probability of delta ray production. Many processes lead to the identification of electrons.

    a) The trapping energy of the chamber is 2 GeV/c so an electron of less than 1 GeV will curl up in the chamber.

    b) A medium energy electron may have a bremsstrahlung with a possible conversion of the radiated gamma ray.

    c) A high energy track may appear as a triplet which is a bremsstrahlung gamma ray superimposed on the straight track.

    d) An electron can undergo an elastic collision and create a δ ray whose energy is larger than that which a hadron can produce.

The electon detection efficiency was determined empirically from γ ray ($e^\pm$ pairs). If one track of a pair is identified as a γ ray the opposite track was examined to see if it could be identified as an

electron. The average detection efficiency of 1/3 for
high energy electrons is consistant with the need for
the electron to go 2 radiation lengths for detection
and the reduced fiducial volume which that would require.

SCANNING PROCEDURE

All of the film was scanned by professional scanners. Every event found was examined by a physicist
for e^+ or e^- candidates. This was the only criteria for further examination of the event. All of
the e^+, e^- candidates were carefully edited, measured,
and analyzed. They were examined for evidence of
strange particles, possible backgrounds such as
comptons, asymmetric dalitz pairs, and decays leading to an $e^\pm$ in the final state.

THE DATA

The results of an early analysis of a partial
scan of the film fall in two samples. For the first
sample (Table 1) 1500 events were completely examined
and backgrounds estimated. The results were 4
events with an e^+ and a K^o_s associated with the main vertex. The remarkable feature of these events is that
each event had a strange particle associated with it.
In addition they were all K^o_s and not a Λ.

BACKGROUND

1) Asymmetric Dalitz Pairs

The reaction
$$\pi^o \rightarrow e^+ e^- \gamma$$

where the e^- is of such low energy so as not to be
detected, represents the major source of background.
We estimate for an $E_{e^-} < 5$ MeV compared to a total
e^+e^- energy of 2 GeV, the average energy of a single
e^+, that the probability for such an aysmmetric pair
is $\sim 3.5 \times 10^{-3}$ per Dalitz pair. This implies <0.2
events in our sample.

2) Leptonic decays of strange particles into
e^+ where the decay is not detected. Such a reaction
would be
$$\nu N \rightarrow \mu^- K^+ K^o X$$
$$\hookrightarrow e^+ \pi^o \nu$$

We estimate <0.1 events in our sample.

3) Leptonic decays of K^o_L at the vertex

$$K^o_L \to e^+ \pi^- \nu$$

We estimate $< 5 \times 10^{-3}$ events

4)
$$K^+ \to e^+ \nu$$
$$\pi^+ \to \mu^+ \nu$$

We estimate $< 5 \times 10^{-4}$ events

5) Neutron interactions near the main vertex. We estimate $< 2 \times 10^{-4}$ events.

6) Anti-electron neutrino interactions

$$\bar{\nu}_e N \to e^+ + X$$

However for such an event we expect the event to be at low y because anti-neutrino reactions are expected to fall off in y as $(1-y)^2$ where the e^+ is interpreted as the primary lepton. For our events y > 0.8 when interpreted as $\bar{\nu}_e$ events.

We estimate $< 4 \times 10^{-3}$ events.

In conclusion we expect <.3 events as background where we observe 4 events. We feel we have been very generous in these estimates and that the background is smaller. It should be emphasized that the dominate source of background is asymmetric Dalitz pairs. Any contribution from this background would contribute e^+ and e^- with equal probablity so we expect to find background $\mu^- e^-$ events while none are observed. In addition the other major contributers to background are expected to contribute e^- as well as e^+ condidates which have not been observed. The use of high resolution photographic film to record these events aids in the elimination of backgrounds such as close in Vee's. The enlargement of the vertex on the film and careful examination can only be done with a bubble chamber.

For the second sample (Table 2) three more events were found. These events were all unique. The first event, C_1 was found to have a K^o_S and a Λ pointing to a recoil. These fit the reaction

$$K^o_L + p \to \pi^+ + \Lambda$$

We consider this event as the first with 2 strange
particles detected. The second event has a Vee near
the main vertex which fits a K_S^o. There are 2 Vee's
several K_S^o lifetimes away which we interpret as the
3 body decays of K_L^o. In addition 3 gammas point to
a common origin near the main vertex. These we interpret as the neutral decay of a K_S^o

$$K_S^o \rightarrow \pi^o \; \pi^o$$
$$ \downarrow \;\; \downarrow \gamma\gamma$$
$$ \hookrightarrow \gamma\gamma$$

The third event W5 has no Vee and is of low energy.
It is only 50 cm from the wall of the chamber and is
the most difficult event to examine in detail.
These three events come from a sample of about 1000
events so the background can be approximately
scaled. However they have not been as accurately estimated.

THE ANALYSIS

Some of the relevant kinematic quantities are
presented in Table 3. Note that the average ratio of
the $P_{\mu^-}/P_{e^+} \approx 5$. This is larger than the limit estimated by Pais and Treiman[3] for the production of a
heavy lepton. On a $2M\nu$ vs. Q^2 plot (Fig. 1) the
events scatter although they tend to be at larger
W^2 than for a typical event of 100 events from
this experiment. On an x vs y plot (Fig. 2) we
find that events tend to cluster at lower values of
x and y > 0.4. In addition the variable $u = y(1-x) \approx$
$\frac{W^2}{S}$ is larger than 0.4. The exception to these
general observations is the third event W3. This
event is a 2 prong with K_S^o. It has 4 GeV of missing transverse momentum. We expect the total
missing momentum to be larger than this. In
both figures 1 and 2 the change in the kinematic
quantities are indicated by the line associated with
this event. A recent analysis by Barger and Phillips[4]
based on a model in which a new particle is produced
whose leptonic decay proceeds in much the same way as
K_{e3} decays has predicted the distribution in trans-

verse momentum of the wrong sign lepton with respect to the plane defined through the neutrino and muon. Figure 3 shows the result of the calculation along with the data. We see that a mass for the particle of less than or near 2 GeV is favored.

SUMMARY AND IMPLICATIONS

A brief summary of the results of the few events we have observed lead to several important conclusions.
    1) the events do not appear to be due to $\bar{\nu}_e$
    2) the backgrounds appear to be small
    3) there must be missing neutral energy (neutrino) by lepton conservation e.g. W3
    4) of 7 observed events, 6 have neutral K and 2 have multiple K's. No Λ's have been observed.
    5) The rate of production of these events relative to conventional events

$$\frac{\mu^- e^+ X}{\mu^- X} \sim 1\%$$

    The rate is probably between .5% and 3%.
    6) The ratio P_{μ^-}/P_{e^+} is large because the energy of the e^+ tends to be low.

From these observations we conclude
    1) We have observed the leptonic decay a new hadron.
    2) The transverse momentum of the e^+ relative to the (ν, μ) production plane would imply a mass from 1.5 to 2.0 GeV.
    3) The production and/or decay of this object is related to strangeness and it is presumed to be a meson.

It should again be emphasized that scanners only find events and physicists only find $e^{\pm}$ candidates. The Vee's are found later. We are not biased toward finding events with K_S^o. We certainly can not tell a K_S^o from a Λ on the scanning table very well. The detection efficiency for K_S^o is about 50% while for K_L^o it is about 15%. The number of K_S^o and K_L^o found is roughly compatable with their being produced in equal numbers and further implies that each event has from 2 to 3 K associated with it.

FURTHER WORK

We have more film to edit and expect to have a sample of at least 20 events when we finish. We will be looking for e^+ of lower energy. These we have tended to disregard because the backgrounds are so much higher. We will olook at nonleptonic processes by searching for $K^°\pi$ mass distributions. The dynamics of the events may lead to correlations between production and decay planes which can be investigated. The kinematic quantities may display unique features. The multiple Vee events may be produced off the sea and thus would be associated with small x. The identification of any events with an associated Λ is of great interest.

The observation of these events if proved correct has opened an exciting field of investigation and should prove to be of great importance in our understanding of the nature of high energy interactions.

Since the presentation of this talk event W5 has been removed from the sample because the e^+ is compatable with being a dalitz pair when combined with one of the leaving tracks. In addition several more events have been found two of which appear to have an associated Λ.

REFERENCES

1. A. Benvenuti et al., Phys. Rev. Letters 34, 419, 597 (1975), Phys. Rev. Letters 35, 1199 (1975), Phys. Rev. Letters 35, 1203 (1975), Phys. Rev. Letters 35, 1249 (1975).

2. B. Barish et al., Proceedings of the Paris International Conference on Neutrino Interactions, Ecole Polytechnic, p. 131.
   B. Barish, Int'l Conference on Lepton and Photon Interactions at High Energy, Stanford University, August 1975.

3. A. Pais, S. B. Treiman, Phys. Rev. Letters 35, 1206 (1975).

4. V. Barger et al., "Weak Leptonic Decay Distributions for New Particles", University of Wisconsin preprint C00-495 (1975).

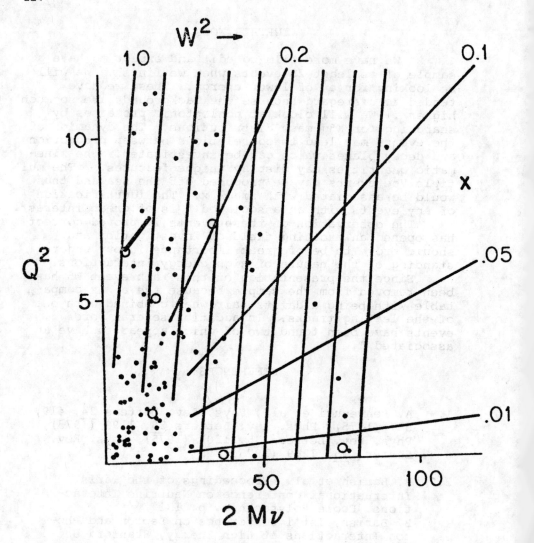

Fig. 1

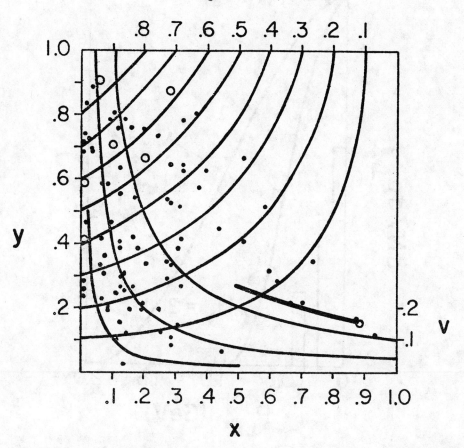

$x = Q^2/2M\nu$ $y = \nu/E = E_h/E_h + E_\mu$

$u = y(1-x) \approx W^2/s$

$v = yx \approx Q^2/s$

Fig. 2

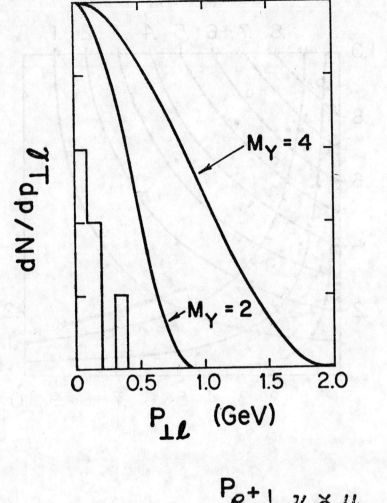

Fig. 3

SEARCH FOR NEW PARTICLE PRODUCTION IN A νH_2 EXPERIMENT
AND IN A $\bar{\nu} H_2$-Ne EXPERIMENT IN THE 15-FOOT
BUBBLE CHAMBER

W. G. Scott
Fermi National Accelerator Laboratory, Batavia, Illinois 60510

ABSTRACT

Searches for evidence for new particle production in the Berkeley-Fermilab-Hawaii-Michigan neutrino hydrogen experiment and in the Fermilab-ITEP-Michigan-Serpukhov antineutrino hydrogen-neon experiment are discussed.

A) AN ANALYSIS OF V-ZERO EVENTS IN THE BERKELEY-FERMILAB-HAWAII-MICHIGAN νH_2 EXPERIMENT (E45)

The Berkeley-Fermilab-Hawaii-Michigan collaboration have analyzed approximately 530 high energy charged current (CC) neutrino events obtained using a broad-band neutrino beam at Fermilab and the 15-Foot bubble chamber filled with hydrogen. In the hydrogen bubble chamber neutral strange particles are detected with high efficiency when they decay to charged particles:

$$\Lambda \to p\pi^-$$
$$K_s^o \to \pi^+\pi^-$$

For new particle searches events with strange particles are of special interest and several aspects of the V-zero sample have been studied in an effort to find evidence for new particle production.
1) A search for exclusive $\Delta S = -\Delta Q$ events

$$\nu p \to \mu^- \Lambda \pi^+ \pi^-$$
$$\mu^- \Lambda \pi^+ \pi^- \pi^+ \pi^-$$
$$\mu^- \Lambda \pi^+ \pi^- \pi^+ \pi^- \pi^+ \pi^-$$
$$\Delta S = -\Delta Q$$

$$\nu p \to \mu^- K_s^o p \pi^+$$
$$\mu^- K_s^o p \pi^+ \pi^+ \pi^-$$
$$\mu^- K_s^o p \pi^+ \pi^+ \pi^- \pi^+ \pi^-$$
possible
$\Delta S = -\Delta Q$

similar to the event reported by Cazzoli et al.,[1] yields no fits.
2) The upper limit on the relative yield of inclusive $\Delta S = -\Delta Q$ production with a Λ, obtained by using all Λ and ΛK_s^o events, is 3.6% at 90% confidence.[2]
3) Inclusive distributions plotted for events with strange particles in this experiment are similar to the distributions plotted for all CC-events.[2] The rate of strange particle production in this experiment appears to be similar to the rate observed in electroproduction and photoproduction experiments.

4) At the New York APS Meeting[3] the results of a systematic study of the following semi-inclusive mass plots were presented:

$\nu p \to \mu^-(\Lambda\pi^+)X$ \qquad\qquad $\nu p \to \mu^-(K_s^o p)X$
$\to \mu^-(\Lambda\pi^-)X$
$\to \mu^-(\Lambda\pi^+\pi^+)X$ \qquad\qquad $\to \mu^-(K_s^o p\pi^+)X$
$\to \mu^-(\Lambda\pi^+\pi^-)X$ \qquad\qquad $\to \mu^-(K_s^o p\pi^-)X$
$\to \mu^-(\Lambda\pi^+\pi^+\pi^+)X$ \qquad\qquad $\to \mu^-(K_s^o p\pi^+\pi^+)X$
$\to \mu^-(\Lambda\pi^+\pi^+\pi^-)X$ \qquad\qquad $\to \mu^-(K_s^o p\pi^+\pi^-)X$
$\to \mu^-(\Lambda\pi^+\pi^+\pi^+\pi^-)X$ \qquad\qquad $\to \mu^-(K_s^o p\pi^+\pi^+\pi^-)X$

$\nu p \to \mu^-(K_s^o \pi^+)X$
$\to \mu^-(K_s^o \pi^-)X$
$\to \mu^-(K_s^o \pi^+\pi^+)X$
$\to \mu^-(K_s^o \pi^+\pi^-)X$
$\to \mu^-(K_s^o \pi^+\pi^+\pi^+)X$
$\to \mu^-(K_s^o \pi^+\pi^+\pi^-)X$
$\to \mu^-(K_s^o \pi^+\pi^+\pi^+\pi^+)X$

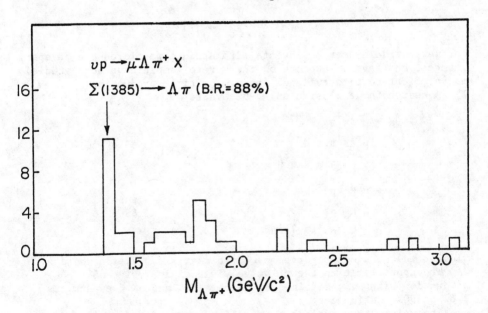

Fig. 1. The $\Lambda\pi^+$ mass plot. The events are plotted in 50 MeV bins.

Note that in general the charged tracks are not identified and each track has to be considered for each possible mass hypothesis in turn. There are 18 Λ and 18 K_s^o in the 3, 5 and 7-prong events. Figure 1 shows the mass plot obtained for the combination $\Lambda\pi^+$. There is a clear $\Sigma(1385)$ signal. A search for a new narrow meson resonance M, or a baryon resonance B produced in the reactions

$$\nu p \to \mu^- M X$$
$$ \hookrightarrow K_s^o + \pi^{\pm}\text{'s}$$

$$\nu p \to \mu^- B X$$
$$ \hookrightarrow \Lambda + \pi^{\pm}\text{'s}$$

$$ K_s^o p + \pi^{\pm}\text{'s}$$

yields the following approximate upper limits on the cross section x {branching ratio} (based on no signal $\gtrsim 7$ events above background) assuming a width less than or on the order of 50 MeV.

$$\sigma_M \times BR(M \to K^o + \pi^{\pm}\text{'s}) \lesssim 4\% \times \sigma_{tot}(CC)$$
$$\sigma_M \times BR(M \to K_s^o + \pi^{\pm}\text{'s}) \lesssim 2\% \times \sigma_{tot}(CC)$$
$$\sigma_B \times BR(B \to \Lambda + \pi^{\pm}\text{'s}) \lesssim 2\% \times \sigma_{tot}(CC)$$
$$\sigma_B \times BR(B \to pK^o + \pi^{\pm}\text{'s}) \lesssim 4\% \times \sigma_{tot}(CC)$$
$$\sigma_B \times BR(B \to pK_s^o + \pi^{\pm}\text{'s}) \lesssim 2\% \times \sigma_{tot}(CC)$$

B) SEARCH FOR "Y-ANOMALIES" IN THE FERMILAB-ITEP-MICHIGAN-SERPUKHOV $\bar{\nu}H_2Ne$ EXPERIMENT (E180)

New data on antineutrino scattering has recently become available from the Fermilab-ITEP-Michigan-Serpukhov antineutrino hydrogen-neon collaboration. The data are based on an exposure of approximately 50000 pictures taken in May 1975 using the 15-Foot bubble chamber filled with a light (21% atomic) hydrogen-neon mixture. The experiment uses a broad-band horn focussed antineutrino beam. An absorbtive plug downstream of the target is used to suppress neutrino contamination. The proton energy was 300 GeV and the mean proton intensity on the target was ~0.8-0.9 x 10^{13} protons per pulse. The experiment uses the Hawaii-Berkeley external muon identifier (EMI) consisting of approximately 4 interaction lengths of zinc absorber together with the bubble chamber magnet coils in the bubble chamber vacuum-vessel followed by approximately 20 m^2 of multiwire proportional chambers.

The film was divided equally between the four laboratories and was scanned for events with total visible momentum along the beam direction p_x more than about 1 GeV. Events with 1-prong only such as for example candidates for the elastic reaction

$$\bar{\nu}p \to \mu^+ + n$$

with an undetected neutron were not included in the scan. In a fiducial volume of 21 m³ approximately 2500 candidates were found of which approximately 1000 have $p_x \gtrsim 10$ GeV.

In order to identify muons in these events all muon candidates are extrapolated to the EMI plane in an attempt to match them with fitted coordinates from the proportional chambers. For this analysis muon candidates which matched in the EMI and for which the hadron confidence level was less than 10% were taken as identified muons. With this criterion the fraction of hadrons hitting the EMI which are misidentified as muons is 10%. Note that not all of the hadrons produced will hit the EMI and the actual fraction of hadrons which are misidentified as muons will be very much less. For energies greater than 10 GeV approximately 60% of events have a μ^+ identified by the EMI.

The energy of the incident antineutrino is estimated by summing the momentum of the muon and the momentum of the hadrons along the beam direction. The momentum of the hadrons has to be corrected to account for undetected energy. Figure 2 shows the mean corrected hadron energy estimated from the mean fractional transverse momentum imbalance plotted versus the hadron energy estimated from the visible hadrons. The data indicate a mean fractional hadron momentum loss in the region of 20-25% together with a constant hadron momentum loss in the region of 0.5-1.0 GeV. Accordingly the estimate for the hadron energy in each event is corrected for undetected energy using the formula given in Fig. 2. For this analysis events with hadron energy ν less than 1 GeV are eliminated from the sample. At small energy transfers the hadron energy resolution is poor and furthermore the event rates are biassed due to the loss of 1-prong events.

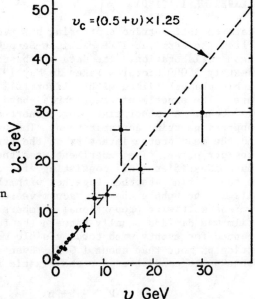

Fig. 2. The correction for missing energy

For the distributions below the event rates shown have been corrected for the geometric acceptance of the EMI. For low momentum muons ($\lesssim 2$ GeV) the acceptance of the EMI is very small and furthermore the effects of random background in the proportional chambers become important. For this reason events where the momentum of the muon is less than 3 GeV are eliminated from the sample. No corrections have been made for losses due to chamber inefficiencies, software inefficiencies or the effects of the muon selection criterion.

The scaling variable y is defined by the relation $y = \nu/E$ where E is the energy of the incident neutrino or antineutrino and ν is the energy transfer to the hadrons in the lab. It is well known that the expected y-distribution for neutrino scattering from spin-½ quarks is flat while the expected y-distribution for antineutrino scattering is proportional to $(1-y)^2$. The reverse is true for scattering from antiquarks. (See Fig. 3).

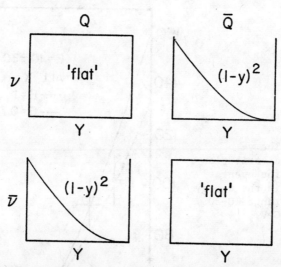

Fig. 3. Schematic y-distribution for ν and $\bar{\nu}$ scattering from spin-½ quarks and antiquarks.

In terms of the quark-parton model the y-distribution measured in antineutrino-nucleon scattering provides a sensitive measure of the relative antiquark content of the nucleon. (See Fig. 4).

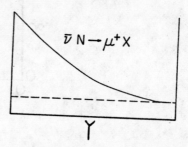

Fig. 4. Schematic y-distribution for inelastic antineutrino nucleon scattering.

The y-distributions shown below have been fitted by the form:

$$dN/dy \propto \{(1-y + y^2/2) - By(1-y/2)\} \qquad (1)$$

In the quark parton model the parameter B is related to the relative antiquark content of the nucleon by the relation:

$$\frac{\bar{Q}}{Q + \bar{Q}} = \frac{1 - B}{2} \qquad (2)$$

where Q and $\bar{Q}$ represent the fractions of the momentum of the nucleon carried by quarks and antiquarks respectively.

Figure 5 a) shows the y-distribution for events in the energy range 10-30 GeV. The solid curve shows a $(1-y)^2$ form and the broken

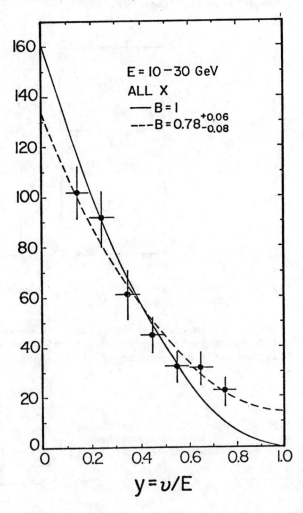

Fig. 5 a). The y-distribution for events in the energy range 10-30 GeV.

curve is a fit to the data. Figure 5 b) shows the same distribution for events in the energy range 30-200 GeV. The two distributions are consistent and the combined distribution is shown in Fig. 5 c). The distributions are inconsistent with a pure $(1-y)^2$ form and a fit to the data gives the result $B = 0.74 \pm 0.06$. In terms of the relative antiquark component of the nucleon this result for B gives
$$\frac{\bar{Q}}{Q + \bar{Q}} = 0.13 \pm 0.03.$$

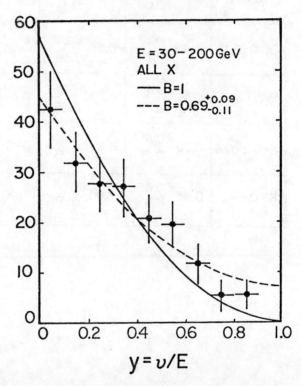

Fig. 5 b). The y-distribution for events in the energy range 30-200 GeV.

Table I shows the value of B fitted for different ranges of the Bjorken scaling variable $x = Q^2/2m\nu$, where Q^2 is the square the 4-momentum transfer. Note that the value of B is consistent with unity for $x > 0.4$ and is significantly less than unity at the smaller x values.

TABLE I Table of Fitted B-Values

	E=10-30	E=30-200	E=10-200
ALL X	$0.78^{+0.06}_{-0.08}$	$0.69^{+0.09}_{-0.11}$	$0.74^{+0.06}_{-0.06}$
X=0-0.1	$0.46^{+0.19}_{-0.27}$	$-0.55^{+0.60}_{-1.14}$	$0.19^{+0.21}_{-0.29}$
X=0.1-0.2	$0.78^{+0.13}_{-0.18}$	$0.73^{+0.19}_{-0.28}$	$0.76^{+0.11}_{-0.14}$
X=0.2-0.4	$0.75^{+0.12}_{-0.15}$	$0.84^{+0.09}_{-0.16}$	$0.79^{+0.08}_{-0.11}$
X=0.4-	$0.95^{+0.05}_{-0.09}$	$1.00^{+0.0}_{-0.04}$	$1.00^{+0.0}_{-0.07}$

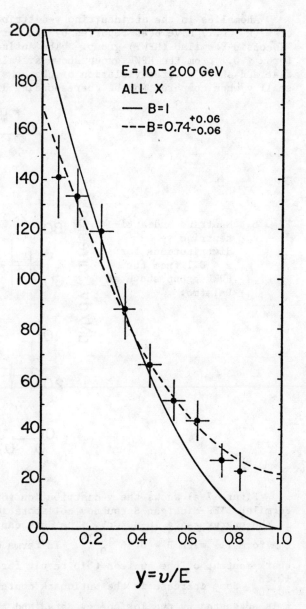

Fig. 5 c). The y-distribution for all events in the energy range 10-200 GeV.

Anomalies in the antineutrino y-distributions for small values of x (x < 0.1) have been reported by the Harvard-Pennsylvania-Wisconsin-Fermilab (HPWF) group.[4] The antineutrino y-distribution for x < 0.1 from the HPWF group shown at Palermo (Fig. 6.) is roughly flat and shows a large violation of charge symmetry invariance at small y when compared to the corresponding distribution for neutrinos.

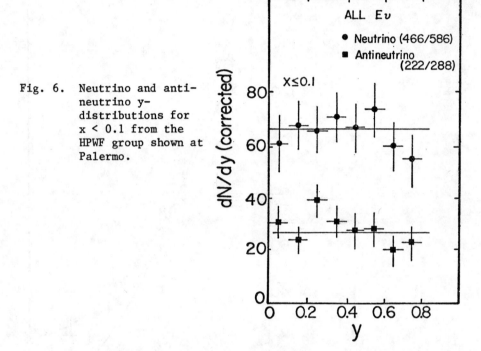

Fig. 6. Neutrino and antineutrino y-distributions for x < 0.1 from the HPWF group shown at Palermo.

Figure 7 a) shows the y-distribution for x < 0.1 from the Fermilab-ITEP-Michigan-Serpukhov collaboration plotted for events in the energy range 10-30 GeV. The data can be fitted nicely by the form (1) with $B = 0.46 ^{+0.19}_{-0.27}$. In terms of the relative antiquark content of the nucleon this result for B gives $\frac{\bar{Q}}{Q + \bar{Q}} = 0.27 ^{+0.13}_{-0.10}$. Some increase in the antiquark content at small x is suggested by the low energy data[5] and there is no evidence for any anomaly in the 10-30 GeV data.

Figure 7 b) shows the same distribution for the energy range 30-200 GeV. Also shown on the same graph are the corresponding data from the HPWF group normalized over the region y < 0.8 The agreement between the two sets of data is quite striking. A fit to the bubble chamber data (broken curve) gives the result $B = -0.55 ^{+0.60}_{-1.14}$. The solid curve is for B = 0.46 from the fit to the 10-30 GeV data normalized using the data for x > 0.1. While there is some

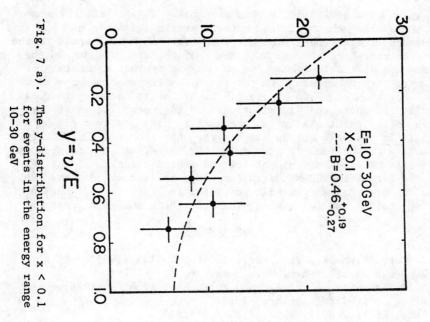

Fig. 7 a). The y-distribution for x < 0.1 for events in the energy range 10-30 GeV

Fig. 7 b). The y-distribution for x < 0.1 for events in the energy range 30-200 GeV.

indication of a deficit of events at small y in the bubble chamber data it is clear that with the present statistics it is not possible to rule out a distribution of the form (1) with B ~0.5 based on the bubble chamber data alone. Note too that the errors shown in the figures are statistical only and do not yet include estimates for systematic errors arising due to errors in the hadron energy estimate. For the future it is clearly important to investigate possible distortions in the distributions due to uncertainties in the hadron energy particularly for the small x high energy region. A study is in progress to investigate the characteristics of the events at small x and to look in particular for correlations with strange particle production or any possible di-lepton signal.

The Fermilab-ITEP-Michigan-Serpukhov collaboration wishes to thank the Berkeley and Hawaii groups for technical assistance in running the EMI and for making available their analysis programs.

REFERENCES

[1] E. G. Cazzoli et al., Phys. Rev. Lett. 34, 1125 (1975).
[2] J. P. Berge et al., Phys. Rev. Lett. 36, 127 (1976).
[3] F. A. DiBianca et al., Bull. Am. Phys. Soc. A9.6 Jan(1976).
[4] D. Cline invited talk Palermo 1975.
[5] H. Deden et al., Nucl. Phys. B 85 269 (1975).

ψ AND ψ' DECAYS AND INTERMEDIATE STATES:
RESULTS FROM SPEAR, LBL-SLAC

Carl E. Friedberg
Lawrence Berkeley Laboratory, University of California
Berkeley, California 94720

INTRODUCTION

Thank you, Professor Chinowsky. First, since this talk is to be of limited duration, I will pass on to my co-conspirator, Dr. Breidenbach, the exposure of the list of collaborators. I will describe, in detail, some recent improvements in the LBL/SLAC magnetic detector.

One year ago (Fig. 1), we installed additional track chambers in the detectors to improve both the solid angle over which charge

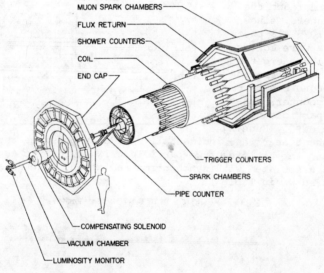

Fig. 1. Exploded view, LBL/SLAC magnetic detector.

particles are detected, and the muon detection system. Last summer, the UCLA group added an antineutron experiment on both sides of the detector. These new facilities are shown in Figs. 2, 3, and 4. The muon tower of power, or towering inferno, has three levels of penetration, each corresponding to about 30 cm of iron (240 MeV/level). The proportional chambers subtend polar angles down to about $32°$, which matches the innermost of the cylinder spark chambers. At the south end of the detector I have installed the four layers of r-φ readout "end cap" chambers, which subtend polar angles as small as

≈ 16°, for three-point tracks. Thus, we can now detect the presence of charged particles over 90% of the solid angle, and we even have some momentum resolution over all of that. Of course, the trigger still requires two charged particles into the central 65% of the solid angle, as before. However, there is now a relaxed trigger requirement which was made possible by the successful operation of the MWPC. Whereas previously two energetic charged particles were required to penetrate the coil and reach the shower counters, since September 1975 all the ψ and ψ' data were collected requiring one energetic particle to traverse the coil, a second one to reach a trigger counter, and at least one particle to strike the MWPC. Running

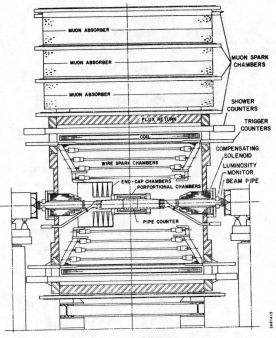

Fig. 2. Detector - side view.

at the <u>highest</u> energies, incidentally, is only possible requiring the MWPC and the "stiffer" two-particle trigger, due to an enormous background associated with synchrotron radiation. The data we have amassed is shown in Table I.

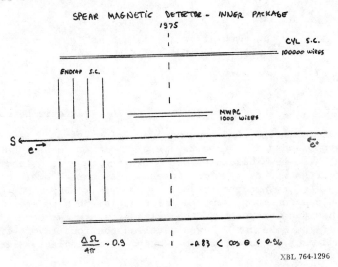

Fig. 3. Detector inner package - top view.

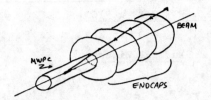

SPEAR LBL/SLAC DETECTOR
INNER CYLINDRICAL MWPC AND ENDCAP CHAMBERS

XBL 764-1298

Fig. 4. Detector - end cap chambers.

Table I. Magnetic Detector Data Collected

SPEAR I	($\sqrt{s}$ energies in GeV)
April '73 ↓ June '74	2.4 - 5.0 (in 0.1 or 0.2 steps)
	3.0, 3.8, 4.8 "big block"

SPEAR II	
July '74 ↓ July '75	5.0 - 8.0 (in 0.05 steps)
	3.1 - 8.0 (in 0.002 - 0.003 steps- scan)
	3.100 - ψ
	3.700 - ψ'
	4.100 big block
	6.200
	7.800

Fall '75 ↓ Spring '76	4.400 region - fine scans
	$\psi, \psi' - \approx$ 300 K events
	7.4 - fine scans
	6.0 - fine scans

$\psi(3100)$

As everyone here knows, in November of 1974 we and MIT/BNL announced the discovery of a very narrow resonance. Now, I will discuss primarily new results. I can recommend that you study George Trilling's SLAC Summer Institute lectures (LBL-4276), and the relevant sessions reported in the Proceedings of the 1975 International Symposium on Lepton and Photon Interactions at High Energies. In Fig. 5, I have summarized the salient features of ψ production, as observed via electron-positron annihilation, and subsequent decays. It has $J^{PC} = 1^{--}$, odd G-parity, and isospin zero. Other important parameters are summarized in Tables II and III. We have examined various predictions of SU_3 to determine if the $\psi(3100)$ is an SU_3 singlet. In principle, the $I = 0$ ψ may be an SU_3 singlet; if so, the following decay modes are forbidden:

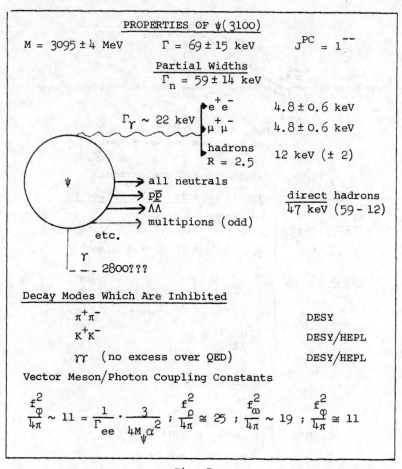

Fig. 5

Decay Mode	90% Confidence Limit
$K_L^0 K_S^0$	< 0.02%
$K^*(890)\bar{K}^*(890)$	< 0.06%
$KK^*(1420)$	< 0.19%
$K^*(1420)\bar{K}^*(1420)$	< 0.18%

Allowed and observed decays are:

Decay Mode	SU_3 Predicted Relative Rate	Observed Branching Ratio
$K^0 K^*(892)$	1	0.24 ± 0.05%
$K^\pm K^{\mp *}(892)$	1	0.34 ± 0.07%
$K^{*0}(892)K^{*0}(1420)$	-	0.37 ± 0.1%
$\rho\pi$	3/2	1.3 ± 0.3%

Table II. Properties of the ψ Particle as Obtained from Fit to Cross Sections σ_{HAD}, $\sigma_{\mu\mu}$ and σ_{ee}.

	ψ_{3095}	ψ_{3684}
Mass	3.095 ± 0.004 GeV	3.684 ± 0.005 GeV
J^{PC}	1^{--}	1^{--}
$\Gamma_e = \Gamma_\mu$	4.8 ± 0.6 keV	2.2 ± 0.3 keV
Γ_H	59 ± 14 keV	220 ± 56 keV
Γ	69 ± 15 keV	225 ± 56 keV
Γ_e/Γ	0.069 ± 0.009	0.0097 ± 0.0016
Γ_H/Γ	0.86 ± 0.02	0.981 ± 0.003
Γ_μ/Γ_e	1.00 ± 0.05	0.89 ± 0.16

Errors accounted for:
  (a) statistical
  (b) 15% uncertainty on hadron efficiency
  (c) 100 keV setting error in E_{cm}
  (d) 2% point-to-point errors, uncorrelated
  (e) 3% luminosity normalization

Thus, there is a factor 2-3 discrepancy between ρπ and the K channels which is not yet understood. Work is still continuing in this area.

We also have some progress to report in the area of baryonic decay modes. We have already reported the branching ratio of $\psi \to p\bar{p}$ = 0.2 ± 0.04% based on ≈ 100 events. These events are identified kinematically, as shown in Fig. 6. We can also use time of flight to identify slower protons and antiprotons, as shown in Fig. 7. Using this TOF identification, we have found that $\psi \to \Lambda\bar{\Lambda}/\psi \to$ all ~ 0.16 ± 0.08%. We have no branching ratio to report yet but looking ahead to Fig. 19, you can see a very clean signal for $\psi \to \pi^+\pi^-p\bar{p}$, about 100 events. We also have a good signal for $\psi \to p\bar{p}\eta$ and $p\bar{p}\omega$.

I will come back later to two more topics related to ψ(3100): branching ratio for $\psi \to$ (all neutral particles) and the search for $\psi \to \gamma + X(2800)$.

ψ'(3700)

In Fig. 8 you will see two events (a and b) which are candidates for $\psi'(3700) \to \Xi^-\bar{\Xi}^-$ (Λπ), and one (c) which is for $\psi'(3700) \to \Xi^-\bar{\Xi}^- X^0$. Figure 9 shows the diplot of $\Lambda\pi^-$ vs $\bar{\Lambda}\pi^+$ effective masses, with a clean peak at the Ξ. We will have to wait to see if those events contain any $\Omega^-\bar{\Omega}^-$ events!

Next, I have begun a search for a heavy, charge 1 object ($H^{\mp}$) in the ψ' events. I do this by taking a sample of events which have total charge 0, missing mass less than 300 MeV/c². I then remove the slowest positive-charged particle, and plot the effective mass of the remaining charged particles. The same is then done removing the slowest negatively charged particle. I thus can search for:

Table III. Decay Modes of the $\psi(3095)$

Mode	Branching Ratio (%)	# of Events Observed	Comments
e^+e^-	6.9 ± 0.9	ca 2000	
$\mu^+\mu^-$	6.9 ± 0.9	ca 2000	
$\rho\pi$	1.3 ± 0.3	153 ± 13	> 70% of $\pi^+\pi^-\pi^0$
$2\pi^+ 2\pi^-$	0.4 ± 0.1	76 ± 9	
$2\pi^+ 2\pi^- \pi^0$	4.0 ± 1.0	675 ± 40	20% $\omega\pi^+\pi^-$; 30% $\rho\pi\pi\pi$
$3\pi^+ 3\pi^-$	0.4 ± 0.2	32 ± 7	
$3\pi^+ 3\pi^- \pi^0$	2.9 ± 0.7	181 ± 26	
$4\pi^+ 4\pi^- \pi^0$	0.9 ± 0.3	13 ± 4	
$\pi^+\pi^- p\bar{p}$		≈ 100	
$\pi^+\pi^- K^+K^-$	0.4 ± 0.2	83 ± 18	not including $K^*(892)K^*(1420)$
$2\pi^+ 2\pi^- K^+K^-$	0.3 ± 0.1		
$K_S K_L$	< 0.02	≤ 1	90% C.L.
$K^0 K^{0*}(892)$	0.24 ± 0.05	57 ± 12	
$K^{\pm} K^{\mp *}(892)$	0.31 ± 0.07	87 ± 19	
$K^0 K^{0*}(1420)$	< 0.19	≤ 3	90% C.L.
$K^{\pm} K^{\mp *}(1420)$	< 0.19	≤ 3	90% C.L.
$K^{*0}(892)\bar{K}^{*0}(892)$	< 0.06	≤ 3	90% C.L.
$K^{*0}(1420)\bar{K}^{*0}(1420)$	< 0.18	≤ 3	90% C.L.
$K^{*0}(892)K^{*0}(1420)$	0.37 ± 0.10	30 ± 7	
$p\bar{p}$	0.21 ± 0.04	105 ± 11	assuming $f(\theta) \sim 1 + \cos^2\theta$
$\Lambda\bar{\Lambda}$	0.16 ± 0.08	19 ± 5	
$p\bar{p}\eta$			
$p\bar{p}\omega$			
$p\bar{p}\pi^0$			
$np\pi^-$	0.37 ± 0.19	87 ± 30	
$\bar{p}n\pi^+$			

$$\psi' \to (\pi^{\pm}\pi^0) H^{\mp}$$

with $H^{\mp} \to \pi^{\mp}\pi^{\mp}\pi^{\pm}, \pi^{\mp}\pi^{\mp}\pi^{\mp}\pi^{\pm}\pi^{\pm}$, etc. I find no events (see Fig. 10). This leads to

$$BR(\psi' \to (\pi^{\pm}\pi^0)H^{\mp}) \times BR(H^{\mp} \to \text{charge 1 state})$$
$$\lesssim 5 \times 10^{-3}, \quad 90\% \text{ C.L.}$$

This work is still in progress.

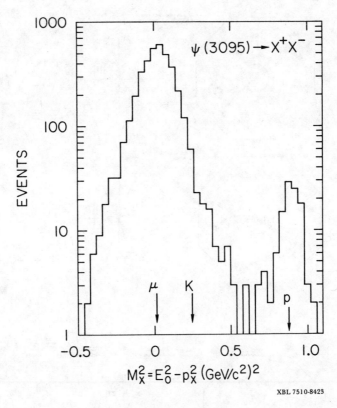

Fig. 6. Mass-squared distribution; protons and muons are cleanly separated.

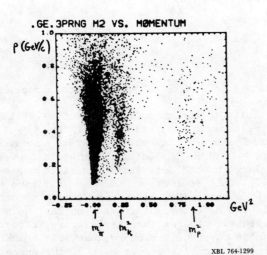

Fig. 7. Mass-squared vs momentum (note antiprotons, kaon, and pin bands).

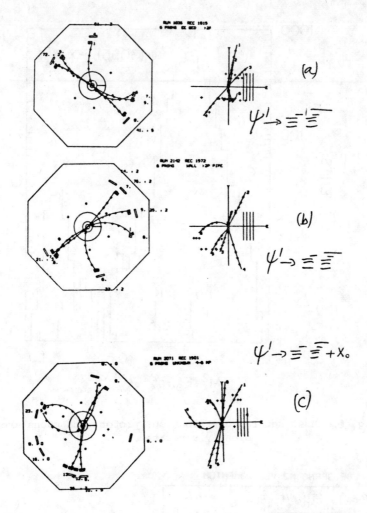

Fig. 8. Candidate events: $\psi' \rightarrow \Xi^-\overline{\Xi}^+$, computer reconstruction.

Now I will return briefly to the topic of all neutral decays of the ψ. This is, of course, nice to know, since (1) $\Gamma_e/\Gamma_{true} = 1 - B_n$, $\Gamma/\Gamma_{true} = (1 - B_n)^2$ where Γ is the total hadronic width we observe and $B_n = (\psi \rightarrow$ unseen$)/(\psi \rightarrow$ all$)$. In addition, it would be very exciting if there were a large unseen decay mode, an extreme example of which might be $\psi \rightarrow \nu\overline{\nu}$. This mode can be studied utilizing the well known cascade decay (see Table IV), $\psi' \rightarrow \pi^+\pi^-\psi$, with no other charged tracks observed in the detector. I remind you that we now have $\approx 10^5$ ψ' events collected with the less restrictive trigger mode, and the 90% solid angle coverage for charge particle tracking; these are crucial to making an accurate measurement of the unseen modes of ψ.

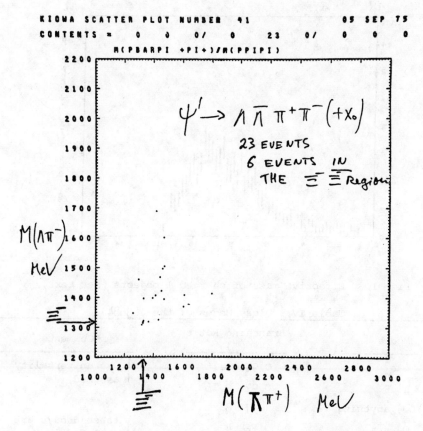

Fig. 9. Mass ($\Lambda\pi$) system vs mass ($\overline{\Lambda\pi}$) system.

Since both of the cascade pions are relatively slow, the relaxed trigger improves our triggering efficiency for $\psi' \to \pi^+\pi^-$ (ψ unseen) about a factor of 10! And since we can now track charge particles over 90% of the solid angle, very few charged decays will be able to simulate neutral decays by missing the detector. In fact, the only significant two particle charged decays faking the all neutral decays would be:

Mode	B.R. of Faked Decays
$\psi \to \mu^+\mu^-$ or e^+e^-	$\lesssim 0.4\%$
$\to \pi^+\pi^-$ + neutrals; e.g., $\rho^0\pi^0$	$\lesssim 0.1\%$
$\pi^+\pi^-\pi^0\pi^0$	$\lesssim 0.1\%$
$\to$ (4 or more charged particles)	$\lesssim 10^{-4}$

Figure 11 shows a typical
$$\psi' \to \pi^+\pi^-\psi \to \text{unseen event.}$$

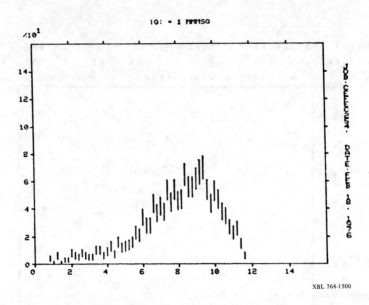

Fig. 10. Effective mass of charged 1 objects (see text).

Table IV. Decay Modes of the ψ(3684).

Modes	Branching Ratio (%)	Comments
e^+e^-	0.97 ± 0.16	μ-e universality assumed
$\mu^+\mu^-$	0.97 ± 0.16	
ψ(3100) anything	57 ± 8	
ψ(3100) $\pi^+\pi^-$	32 ± 4	these decays are included in the fraction for ψ + anything
ψ(3100) η^o	4.3 ± 0.8	
ψ(3100) γγ	< 6.6*	via an intermediate state
$\rho^o \pi^o$	< 0.1*	
$2\pi^+ 2\pi^- \pi^o$	< 0.7*	
$p\bar{p}$	< 0.03*	
$\Xi^- \bar{\Xi}^-$		observed

*90% confidence limit based on a preliminary analysis

Figure 12 shows the two-prong mass-squared spectrum from which cascade decay candidates are selected. There is ≈ 10% background under the cascade peak. There are currently ≈ 1,000 candidates with no extra tracks in ≈ 100,000 events, leading to all neutral B.R. ≈ (2 ± 2)%.

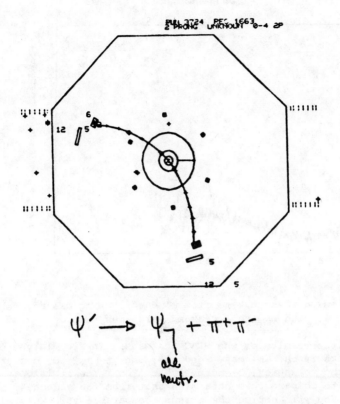

XBL 764-1301

Fig. 11. Candidate event: $\psi' \to \pi^+\pi^-\psi$, $\psi \to$ all neutrals.

SPECTROSCOPY

We all enjoyed the excellent talk given Monday afternoon by T. Appelquist of Yale. Let us recall the bookkeeping on ψ' decays:

$\psi' \to \psi(3100)$ + anything $57 \pm 8\%$

$\psi' \to$ leptons $2 \pm 0.2\%$

Thus, approximately 40% of the ψ' decays (88 keV) are unexplained. Can we shed any new light on these decays? No significant hadronic decay modes have been observed (see Table IV again).

Of course, we do have a proposed spectroscopy, charmonium. And it would be exciting if that is where the missing ψ' decays have gone. In Fig. 13, I have combined two figures from Harari's talk last summer. Perhaps Dr. Breidenbach will make some remarks about the states above ψ'. I will now consider the cascade decays of ψ' and ψ into various intermediate states.

At this time, I must present the results of two different analyses of our data for the decay $\psi' \to \gamma X$, $X \to$ (hadrons), as the Berkeley and SLAC teams have slightly different results. First, I

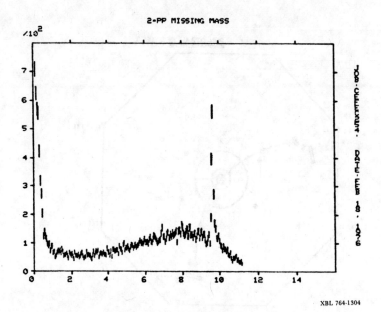

Fig. 12. Missing mass in $\pi^+\pi^-$ from ψ' data showing cascade decays into ψ, with no other charged particles seen.

will present the results of the SLAC analysis. In Fig. 14 you can see three peaks in the $4\pi^\pm$ mass spectrum; one at ≈ 3400, one at ≈ 3550, and the "elastic" (0-energy γ) at 3700. Now it is very illuminating to compare this data with that from the ψ decays, shown in Fig. 15. You will notice there is no comparable structure; in fact a crude limit on

$$BR(\psi \to \gamma X(2800); X(2800) \to 4\pi^\pm) < 1.4 \times 10^{-3} \ .$$

Could there be a valley between 2.9 and 3.0? In Fig. 16, we see candidates for $\psi' \to \gamma X$, $X \to \pi^+\pi^-$ or $X \to K^+K^-$; thus leading to a branching ratio:

$$\psi' \to \gamma X(3470), \quad X(3420) \to \genfrac{}{}{0pt}{}{\pi^+\pi^-}{\text{or } K^+K^-} = 6 \pm 2 \times 10^{-4} \ .$$

There is no clear signal yet for $\psi \to 2800$ in our data. In Figs. 17-20 we see the comparable results from LBL for analysis of various cascade states. There is clear evidence for the X(3420), and for the 3.55; but the case for the 3.45 may be less convincing. Note that in Figs. 17, 18 and 19 the lower portion is the same mass spectrum as observed in ψ decays. In Fig. 20, however, the lower plot is the 6π spectrum with no $\psi\pi\pi$ cascade cuts; this more than doubles the signal. However, note that the 3.49 and 3.55 have moved. Clearly, much more work is needed here.

Finally, I would like to show you the results on inclusive photon spectra being done by Scott Whitaker. In Fig. 21, you will see our data for

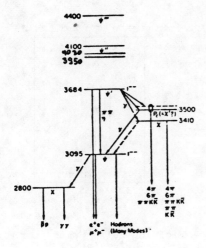

Fig. 13. "Charmonium" levels — predicted (lower) and observed (upper).

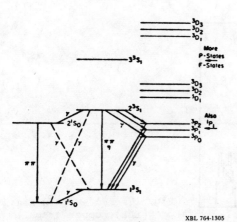

Fig. 14. $\psi' \to \gamma \pi^+\pi^+\pi^-\pi^-$ (SLAC).

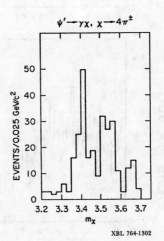

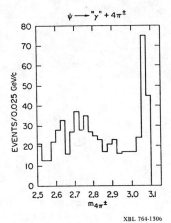

Fig. 15. $\psi \to \gamma \pi^+ \pi^- K^+ K^-$ (SLAC).

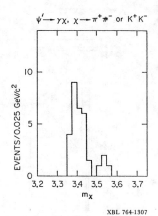

Fig. 16. $\psi' \to \gamma \pi^+ \pi^-$ or $\gamma K^+ K^-$ (SLAC).

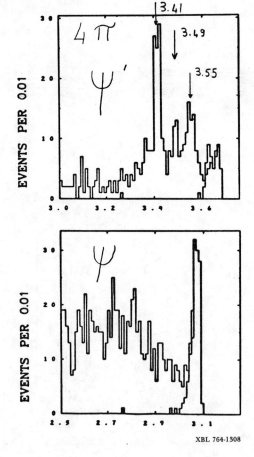

Fig. 17. $\psi' \to \gamma \pi^+ \pi^- \pi^+ \pi^-$ (LBL), upper graph. $\psi \to \gamma \pi^+ \pi^- \pi^+ \pi^-$ (LBL), lower graph.

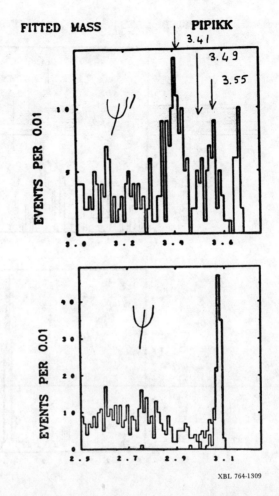

Fig. 18. $\psi' \to \gamma\pi^+\pi^-K^+K^-$ (LBL), upper graph.
$\psi \to \gamma\pi^+\pi^-K^+K^-$ (LBL), lower graph.

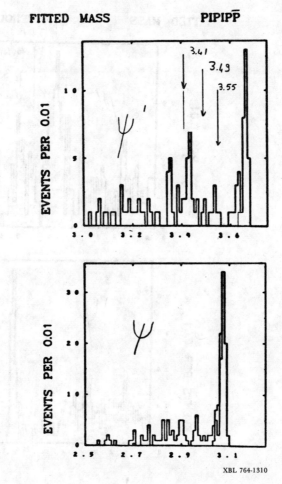

Fig. 19. $\psi' \to \gamma\pi^+\pi^-p\bar{p}$ (LBL), upper graph.
$\psi \to \gamma\pi^+\pi^-p\bar{p}$ (LBL), lower graph.

245

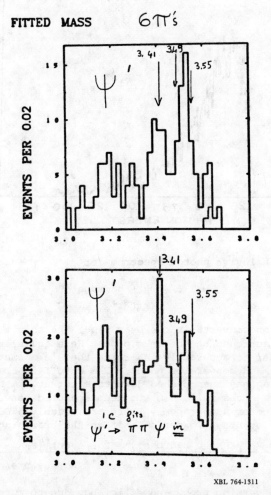

Fig. 20. $\psi' \to \pi^+\pi^+\pi^+\pi^-\pi^-\pi^-$ (LBL), upper graph.
$\psi \to \pi^+\pi^+\pi^+\pi^-\pi^-\pi^-$ (LBL), lower graph, no cascade cuts and including 1C fits.

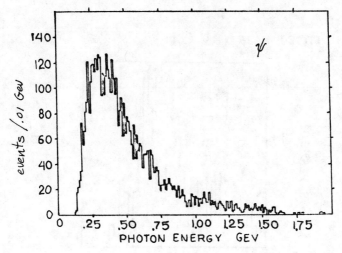

Fig. 21. Inclusive photon spectrum for $\psi \to \gamma$ + hadrons.

$$\psi \to \gamma + \text{hadrons}$$
$$\hookrightarrow e^+ e^-$$

where the photon converts in the beam pipe, the scintillation counters, or the MWPC's surrounding the beam pipe. Here again, the MWPC's are crucial in order to accurately determine the e^+e^- momentum, and the opening angles and conversion point. The cutoff in transverse momentum for tracking has been pushed down to a cut at $p_\perp > 0.055$ GeV/c. We also require $|\cos \theta_\gamma| < 0.6$ to keep accepted conversions well within the detector acceptance. There are roughly 5400 γ's converted from 130,000 ψ events. To set a limit on the branching ratio, we use

$$\text{BR:} \quad \frac{\Gamma(\psi \to \gamma X[X \to \text{hadrons}])}{\Gamma(\psi \to \text{all})} = \frac{\text{No. events}/[\epsilon_{th}^s \cdot \alpha \cdot p \cdot R]}{\text{No. events}/\epsilon_{th}^{all}}$$

$\alpha = 0.009$ = conversion prob, and efficiency due to cuts on e^+e^-, etc.; $p = 0.86$ = post cuts; $R = 0.8$ = radiative corrections; and $\epsilon_{th}^{all}/\epsilon_{th}^s \sim 0.8 \pm 0.2$ is the relative triggering efficiency (the electrons have a higher likelihood of penetrating the coil and reaching the shower counters). Then, since no events are seen at $E_\gamma \approx 300$ MeV [for X(2800)] we can include:

$$\text{BR}[\psi \to \gamma X \to \gamma(\text{hadrons})] < 10\%, \quad 90\% \text{ C.L.}$$

The situation for the ψ' can be seen by overlaying Fig. 21 with Fig. 22. A clear peak containing about 200 ± 40 events has an energy of 265 ± 5 MeV. This leads to the result

$$\frac{\psi' \to \gamma X(3420), X \to \text{hadron}}{\psi' \to \text{all}} = \begin{cases} 8 \pm 4\%, & \text{for isotropic } \gamma \text{ distribution} \\ 10 \pm 4\%, & \text{for } (1 + \cos^2 \theta) \gamma \text{ distribution} \end{cases}$$

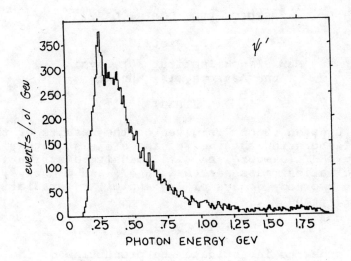

Fig. 22. Inclusive photon spectrum for $\psi' \to \gamma$ + hadrons.

So, another 10% is eaten up!

I would like to thank my collaborators for many of the results which I have presented here today, and the Conference organizers for the excellent job they have done.

This work was supported by the U. S. Energy Research and Development Administration.

RESULTS FROM DORIS, DASP

K. P. Pretzl

Max Planck Institut für Physik
und Astrophysik, München

ABSTRACT

An updated report is given on the results of the DASP-Collaboration obtained at the e^+e^-- storage rings DORIS at DESY (Hamburg) on a.) inclusive distributions, particle ratios, mass searches in J/ψ and ψ' decays, b.) hadronic two body decays of J/ψ and ψ', c.) radiative decays of J/ψ and ψ'.

INTRODUCTION

In order to investigate the properties of the recently discovered J/ψ and ψ' particles most of the data were taken with the DASP (Double Arm Spectrometer) detector at the energies E_{CM} = 3.09, 3.68 and 4.15 GeV.

As already reported at SLAC Conference (Aug. 1975) evidence has been found for a radiative decay of J/ψ into a state X(2.75-2.8 GeV). This has been under further investigation ever since the experimental program at DORIS was resumed after a shut down period in fall 1975. Since data taking was finished just a few weeks before this conference and the data analysis has not yet been completed, no new results on the status of the X-particle can be reported at this conference.

The results of the following topics are presented here:
1.) Inclusive distributions, particle ratios and mass searches in J/ψ and ψ' decay.
2.) Hadronic two body decay of J/ψ and ψ'.
3.) Radiative decays of J/ψ and ψ'.

APPARATUS

The experiment was performed at the e^+e^-- storage rings (DORIS) at DESY (Hamburg) which reach a maximum energy of E_{CM} = 9 GeV. The luminosity at E_{CM} = 3.68 GeV was typically $L = 3.10^{29}$ cm^{-2} sec^{-1} avaraged over a days running time.

The DASP-detector has already been described elsewhere [1]. Therefore only its main features are listed here.

There are two large H-magnets of opposite polarity on either side of the interaction point (see Fig. 1).

Using proportional wire chambers and magnetostrictive wire chambers the direction of charged particles was measured before they enter and after they passed the magnet. The momentum resolution obtained with a magnetic field length of 1.1 Tm was $\Delta p/p = \pm 1.0\%$. Hadrons were identified via time of flight measurement. As shown in Fig. 3 pions, kaons and protons are well separated at all momenta up to the kinematical boundary.

Electrons are distinguished from hadrons via the pulse height information in the shower counters positioned in the outer spectrometer arms. Muons are identified by their range through 70 cm of iron.

The inner detector, covering 70% of 4π, is shown in Fig. 2. It is in a field free region and consists of proportional chambers and units of scintillators, lead, proportional tube chambers and shower counters. With this detector the energy and direction of photons could be determined to a precision of $\Delta E/E = 30\%$ for a photon of 1 GeV and $\Delta\alpha = \pm 2°$.

Since the DASP-spectrometer allows simultaneous measurement of charged particles with good momentum resolution and photons with good angular resolution it is an ideal instrument in the search for new states in the radiative decay of the J/ψ and ψ'.

RESULTS

I   Inclusive distributions, particle ratios, mass searches.

The inclusive spectra were measured by requiring a single charged particle to traverse one of the arms of the spectrometer. Thus the trigger efficiency is independent of the final state and does not introduce systematic uncertainties between pions, kaons and protons.

Fig. 4 a. and b.) show the invariant cross section $E/4\pi p^2 \cdot d\sigma/dp$ versus the particle energy for the decay $J/\psi \to h^{\pm} X^{\mp}$ and $\psi' \to h^{\pm} X^{\mp}$. Both spectra display a similar general behaviour. The cross section for antiprotons* is larger than the cross section for pions and kaons. In the case of $\psi' \to h^{\pm} X^{\mp}$ the pion yield below 0.5 GeV increases faster than in $J/\psi \to h^{\pm} X^{\mp}$. This can be under-

---

* In the following we will always refer to $2\cdot\bar{p}$ rather than $p^{\pm}$ protons and antiprotons, because the antiproton signal is a clear signal coming only from e^+e^- annihilations and is free of background due to beam gas interaction.

stood in view of the fact that the ψ' decays into the J/ψ via the mode $\psi' \to J/\psi + \pi^+\pi^-$ in 36% of the cases.

The invariant cross sections follow an exponential and can be approximated with the formula
$$E/4\pi p \cdot d\sigma/dp \sim e^{-aE}$$
The values of the exponent a, as obtained from a fit to the data, are listed in table I.

Table I Values of the exponent a, as obtained from a fit to the data.

$h^{\pm}$	$J/\psi \to h^{\pm} X^{\mp}$	$\psi' \to h^{\pm} X^{\mp}$
$\pi^{\pm}$	5.9 GeV^{-1}	5.6 GeV^{-1}
$K^{\pm}$	5.5 GeV^{-1}	4.7 GeV^{-1}
$p,\bar{p}$	6.0 GeV^{-1}	6.4 GeV^{-1}

A similar behaviour of the invariant cross section as a function of P_T is observed in hadron hadron collisions where the cross section at small P_T is well reproduced by
$$E \, d\sigma^3/dp^3 \sim e^{-6P_T}$$

The particle fractions as a function of momentum measured in the J/ψ and ψ' decay are plotted in Fig. 5. The relative yield of pions decreases while the relative yield of kaons and protons increases with increasing momenta. In table II the relative particle fractions for particle momenta between 0.5 and 1.3 GeV/c are listed.

Table II Particle fractions observed for momenta between 0.5 and 1.3 GeV/c.

E_{CM} (GeV)	$\pi^{\pm}$ / all charged	$K^{\pm}$ / all charged	$2.\bar{p}$ / all charged
3.1	0.805 ± 0.006	0.140 ± 0.006	0.056 ± 0.005
3.7	0.86 ± 0.02	0.097 ± 0.015	0.050 ± 0.01
4.15	0.84 ± 0.09	0.10 ± 0.05	0.06 ± 0.05

A search for peaks in the missing mass spectra of the system recoiling against the single hadrons has been performed. The color interpretation of the new particles[2] predicts a prominent decay of the ψ' into a pion and a colored ρ_c with a mass around 3.1 GeV. To search for this decay a special run was taken with a reduced magnet current of 300 Ampers to assure a uniform mass acceptance around 3.1 GeV. As shown in Fig. 6 no peak in the measured spectrum was found. We found

$$\frac{\Gamma(\psi' \to \pi^{\pm} \rho_c^{\mp})}{\Gamma(\psi' \to \text{all})} < 5.10^{-2}$$

at the 90% confidence level. No peaks were found in the missing mass spectra for the decay $\psi' \to \pi^{\pm}$ X, $K^{\pm}$ X, $p\bar{X}$, $\bar{p}X$.

A quasi-two body decay of the type $J/\psi \to \pi^{\pm}\rho^{\mp}$, $J/\psi \to K^{\pm} K^{*\mp}$, $J/\psi \to p \bar{p}$ has been observed as a peak in the missing mass spectra in Fig. 7 a.), b.) and c.). The calculated branching ratio for the decay $J/\psi \to \pi\rho$ is

$$\frac{\Gamma(J/\psi \to \pi\rho)}{\Gamma(J/\psi \to \text{all})} = 0.9 \pm 0.3\%$$

assuming $\Gamma\pi\rho = \Gamma\pi^+\rho^- + \Gamma\pi^-\rho^+ + \Gamma\pi^0\rho^0$ and for the decay $J/\psi \to K K^*$ is

$$\frac{\Gamma(J/\psi \to K K^*)}{\Gamma(J/\psi \to \text{all})} = 0.5 \pm 0.2\%$$

assuming $\Gamma_{K K^*} = \Gamma_{K^+K^{*-}} + \Gamma_{K^-K^{*+}} + \Gamma_{K^0\bar{K}^{*0}} + \Gamma_{\bar{K}^0K^{*0}}$

The effective mass spectrum of the p $\bar{p}$ pairs observed in the reaction $J/\psi \to p \bar{p}$ X is plotted in Fig. 8. The sharp peak at $M_{p\bar{p}}$ = 3.09 GeV with a resolution of $\Delta M_{p\bar{p}}$ = 40 MeV (FWHM) suggests a direct decay of $J/\psi \to p \bar{p}$. The angular distribution of the events $J/\psi \to p \bar{p}$ is plotted in Fig. 9. The poor statistic however does not allow any conclusion on the relative contributions of the electric and magnetic nucleon form factors G_E and G_M.

II Hadronic two body decays of J/ψ and ψ'.

In the search for the decay of the resonances into a pair of pions, kaons and antiproton proton the oppositely charged particles were required to be collinear to better than 0.15 rad and to have the correct momentum within ± 50 MeV. In the case of ψ' decay no events were found in any of these three channels. The decay $J/\psi \to \pi^+\pi^-$ has been seen in one event. The branching ratios for the hadronic two body decays of the J/ψ and ψ' are listed in table III.

Table III Branching ratios of the hadronic two body decays of the J/ψ and ψ'.

decay mode	J/ψ B.R.%	J/ψ Γ(KeV)	ψ' B.R.%	ψ' Γ(KeV)
p p̄	0.24 ± 0.05	0.17 ± 0.034	< 0.034	< 0.075
π⁺π⁻	0.004 ± 0.004	0.003 ± 0.003	< 0.04	< 0.09
K⁺K⁻	< 0.021	< 0.014	< 0.13	< 0.29

The upper limits correspond to a 90% confidence level. For the calculated decay rate J/ψ → p p̄ a $1 + \cos^2\theta$ angular distribution was assumed.

The decay of J/ψ → π⁺π⁻ and J/ψ → K⁺K⁻ is strongly suppressed and occurs at a level where it can be compared with the second order radiative decay as scetched in the following diagram

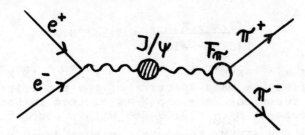

From this diagram we can calculate the apparent pion form factor $F\pi$ using the relation

$$|F\pi|^2 = 4\ \Gamma(J/\psi \to \pi^+\pi^-)/\Gamma(J/\psi \to \mu^+\mu^-)$$

and obtain $|F\pi|^2 = (2.4 \pm 2.4) \cdot 10^{-3}$. If one extrapolates the available data[3] on $\sigma(e^+e^- \to \pi^+\pi^-)$ to E = 3 GeV or uses the rho pole formula the value of $|F\pi|^2$ is of the same order of magnitude. Similarly an upper limit for the kaon form factor of $|F_K|^2 < 1.2 \cdot 10^{-3}$ was deduced from the measured branching ratios.

Fig. 10 shows the ratio $\sigma(e^+e^- \to p\bar{p})/\sigma(e^+e^- \to \mu^+\mu^-)$ as measured in this experiment and compares it with other measurements [4] at and outside the resonances J/ψ and ψ'. At the resonances the ratio is clearly above the value obtained by any extrapolation below the resonances. We conclude that we observe the direct hadronic decay J/ψ → p p̄.

The direct decay into proton antiproton pairs and the strongly suppressed π⁺π⁻ decay mode indicates an isospin and G-parity assignement $I^G = 0^-$ for the J/ψ. The ratio of the πρ to the K K̄* decay suggests that the

J/ψ is a SU_3 singlett.

III Radiative decays of J/ψ and ψ'.

A. Introduction

Among the many models which tried to explain the J/ψ and ψ' states one of the most appealing is the charmonium model[5]. In analogy to positronium the model predicts ortho (spins parallel) and para (spins antiparallel) states and many additional states resulting from spin orbit couplings. The producted states, as shown in Fig. 11, are narrow due to the Zweig-rule suppression of charm decay below threshold.

Since the ortho states carry the quantum numbers of the photon they can be formed directly in e^+e^- collisions, while the para states and P states, for example, can be observed only in the radiative decay of the J/ψ and ψ'. Due to the simultaneous detection capability for photons and charged particles the DASP-spectrometer is a suitable instrument to search for these states.

B. Three photon final states in J/ψ and ψ' decay

When searching for new resonances in the three photon final states the following processes might contribute to the observed event yield:

a.) Quasi two body decays of the type: J/ψ or ψ' → $π^0γ$, η γ, η'γ. There is evidence for a new resonance X decaying into 2 photons at a mass around $M_{γγ}$ = (2.75 - 2.8) GeV.

b.) The direct decay of J/ψ and ψ' into three photons should be very much suppressed and no peaks should appear in the effective mass spectrum for any combination of two photons.

c.) Three photon final states via the QED process

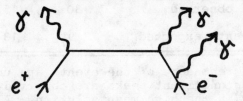

which represents the most serious background source.

Events which had 3 photons in the final state were selected from the total data sample collected with the inner detector. These events were filtered by the following requirements:

a.) The angle between two photons had to be bigger

than 30°. This filter removed all $\pi^0\gamma$ events.
b.) The events had to be coplanar with a coplanarity angle $\Delta\theta \leq 5°$ (see Fig. 12 a.).

Finally the QED background in the remaining sample was reduced by excluding events with an event plane in or close to the beam axis and with a collinearity angle τ smaller than 5° between any combination of two photons (see Fig. 12 b.). Thirty events in the case of J/ψ decay and fifteen events in the case of ψ' decay passed the filter. The effective mass between any combination of two photons was calculated from the measured photon directions and the derived photon energies. The mass resolution ΔM/M at a total CM energy of 3.1 GeV was estimated to be ± 5% for M = 0.5 GeV and ± 1.5% for M = 2.8 GeV.

The three photon events, as measured at the J/ψ and ψ' resonance, are plotted in Fig. 13 a.) and b.) as a function of the lowest and highest effective mass. Also plotted in the diagrams is the QED background which was computed with Monte Carlo methedes*. In the background computation, the events were properly normalized and the photon detection efficiency of the apparatus was taken into account. Cuts identical to the real events were applied. In table IV the number of observed events at and outside the resonances is compared with the number of expected QED events

Table IV Number of observed events at and outside the resonances J/ψ and ψ' compared with the number of expected QED events.

	J/ψ	ψ'	outside
Integrated Luminosity $\int Ldt$ [nb^{-1}]	211	367	118
events observed	30	15	4
QED events expected	9	13	5

At the resonance ψ' the events are uniformly distributed. No significant peaks are observed either in the low mass or in the high mass solution. A clear η peak is observed in the low mass solution at the resonance J/ψ.

---

* The QED events were computed by using the matrix elements for this reaction given by Berend and Gastman[7] as an imput to the Monte Carlo program.

Its width is consistent with the width expected from the experimental resolution. There are two events at the mass of the η'. The clustering of 12 events around the mass of 2.8 GeV in the high mass solution may suggest the existence of a new resonance X. The observed peak is well within the region of uniform mass acceptance which extends out to 3.0 GeV. The width of the peak is consistent with the experimental resolution. The DESY-Heidelberg collaboration[8] also investigated the decay J/ψ → γγγ. They find a clustering of events around 2.8 GeV. Their results are in good agreement with the results of the DASP group.

Since the shut down in the fall 1975 the data sample on the 3 photon final states in J/ψ decay has more than doubled, but the final analysis is not completed at this time. So far the clustering at 2.8 GeV persists.

The branching ratios of the decay J/ψ or ψ' → X(2.8 GeV)γ and other radiative decay modes such as J/ψ or ψ' → γγ, π°γ, ηγ and η'γ are listed in table V.

Table V Branching ratios of the radiative decays of J/ψ and ψ'.

decay mode	J/ψ	ψ'
γγ	$< 3 \cdot 10^{-3}$	$< 8 \cdot 10^{-3}$
π°γ	$< 10^{-2}$	$< 10^{-2}$
ηγ	$(1.3 \pm 0.4) \cdot 10^{-3}$	$< 1.3 \cdot 10^{-3}$
η'γ	$(0.5 \leftrightarrow 4) \cdot 10^{-3}$	$< 1.4 \cdot 10^{-2}$
X(2.8 GeV)γ ↳ γγ	$1.5 \cdot 10^{-4}$	$< 3.7 \cdot 10^{-4}$

C. Other decay modes of J/ψ

The decay J/ψ → η'γ with η' → ρ°γ → π⁺π⁻γ was investigated using the DASP detector. The events were selected by requiring that one of the two pions beeing detected in one of the spectrometer arms and the other, beeing detected in the inner detector, should have an angle between them larger than 50°. Furthermore the angles between the two photons measured in the inner detector had to be larger than 100°. Three events with the proper ρ and η' mass were found after kinematical fitting of the filtered events. The χ^2 was required to be less than 5 for one degree of freedom. The events are shown in Fig. 14 a.) and b.). The measured branching ratio for the J/ψ → η'γ

decay is
$$\frac{\Gamma(J/\psi \to \eta'\gamma)}{\Gamma(J/\psi \to \text{all})} = (0.5 \leftrightarrow 4) \cdot 10^{-3}$$

D. Decay of ψ' via intermediate states P_c

The decay of $\psi' \to J/\psi + \gamma\gamma$ was observed in two nearly independent experiments. In the first experiment the two electrons from the J/ψ decay and the two protons were measured in the inner detector. In the second experiment the muon pair from the J/ψ decay was identified and momentum analysed in the magnetic spectrometer arms and the two photons were measured in the inner detector. Both experiments gave the same absolute rate for the $\gamma\gamma$ transition. Since the photon resolution obtained in the first experiment was not sufficient to separate the various transitions only the second experiment will be discussed here.

Out of the total data sample we found 33 events of the type $\mu\mu$ $\gamma\gamma$ with an effective mass $2.9 < M_{\mu\mu} < 3.2$ GeV. These events still contained background events coming from the decay of $\psi' \to J/\psi + \eta$, $\psi' \to J/\psi \pi^0\pi^0$, where two of the photons are detected. The η background was removed by requiring the effective mass $M_{\gamma\gamma}$ to be less than 520 MeV. A 3c-fit was made to the remaining 29 events in order to remove most of the $\pi^0\pi^0$ background. Nine events passed the fit with a χ^2 of less than 12. The $\pi^0\pi^0$ background in this sample has been estimated to be less than one event.

The events are plotted in Fig. 15 as a function of the energies of the two photons. The events cluster in two groups of 6 events and 2 events each. The first group is centered at $E_{\gamma 1} = (169 \pm 7)$ MeV, $E_{\gamma 2} = (398 \pm 7)$ MeV demonstrating the existence of a new narrow resonance P_c with a mass of $M_{P_c} = (3.507 \pm 0.007)$ MeV or $M_{P_c} = (3.258 \pm 0.007)$ MeV. Since it is not possible to decide, which of the two photons was emitted first, an ambiguous mass has to be assigned to the new resonance P_c.

The second group is centered at $E_{\gamma 1} = (263 \pm 8)$ MeV $E_{\gamma 2} = (315 \pm 8)$ MeV and suggests the existence of a second state P'_c with a mass $M_{P'_c} = (3.407 \pm 0.008)$ GeV or $M_{P'_c} = (3.351 \pm 0.008)$ GeV.

The decay of $P_c \to \pi^+\pi^-$ and $P'_c \to K^+K^-$ have also been observed at an invariant pair mass of repectively 3.515 GeV and 3.439 GeV. The decay into pseudoscalar mesons suggests a natural spin parity assignement to the P_c

states.

The branching ratios as defined

$$\frac{\Gamma(\psi' \to \gamma P_c)}{\Gamma(\psi' \to all)} \cdot \frac{\Gamma(P_c \to f)}{\Gamma(P_c \to all)}$$

are listed for various final states f in table VI.

Table VI Branching ratios for the radiative decay of the ψ' as defined

$$\frac{\Gamma(\psi' \to \gamma P_c \text{ or } \chi)}{\Gamma(\psi' \to all)} \cdot \frac{\Gamma(P_c \text{ or } \chi \to f)}{\Gamma(P_c \text{ or } \chi \to all)}$$

for various final states f.

decay mode f	P_c (DASP) 3507 ± 7 MeV	P_c' (DASP) 3407 ± 8 MeV	χ (LBL-SLAC) 3410 ± 10 MeV	χ (LBL-SLAC) 3530 ± 20 MeV
$\gamma J/\psi$	$4 \cdot 10^{-2}$	$1.5 \cdot 10^{-2}$		
$\pi^+ \pi^-$	$2 \cdot 10^{-4}$	$< 6 \cdot 10^{-4}$	$(6.5 \pm 1.8) \cdot 10^{-4}$	$< 2.7 \cdot 10^{-4}$
$K^+ K^-$	$< 1.2 \cdot 10^{-3}$	$4 \cdot 10^{-4}$		
$4 \pi^{\pm}$			$(1.4 \pm 0.7) \cdot 10^{-3}$	$(2 \pm 1) \cdot 10^{-3}$
6π			10^{-3}	$2 \cdot 10^{-3}$
$\pi^+ \pi^- K^+ K^-$			$7 \cdot 10^{-4}$	$5 \cdot 10^{-4}$

Also shown in this table are the χ-states found by the LBL-SLAC group. Comparing the branching ratio measured for the $\gamma\gamma$ cascade in this experiment with the single monoenergetic photon line obtained by J. W. Simpson et al.[9] one concludes that the decay $P_c \to J/\psi \gamma$ is the major decay mode.

CONCLUSION

A study of the radiative decays of the new resonances lead to the discovery of further narrow states P_c, χ and X with even charge conjugation (see Fig. 16). It is tempting to associate these resonances with the states predicted by the charmonium model. However other models can also explain these new resonances. All we can conclude at this point is, that the J/ψ and the ψ' behave like fermion-antifermion bound states where the fermions possibly carry a new quantum number.

REFERENCES

1. DASP-Collaboration, Phys. Letters 56B, No. 5, p.491 (1975), B. Wiik, Proceedings of the 1975 International Symposium on lepton and photon interactions at high energies (Stanford 1975), p. 69.
2. Recent review: O. W. Greenberg, University of Maryland, technical report No. 75-064.
3. M. Bernardini et al., Phys. Letters 46B, 261 (1973).
4. M. Castellano et al., data quoted by V. Silvestrini; Proceedings of the XVI International Conference on High Energy Physics 1972 (ed. by J. D. Jackson and A. Roberts) Vol. 4, p. 1; M. Conversi et al., Nuovo Cimento 40A, 690 (1965); D. L. Hartill et al., Phys. Rev. 184, 1415 (1969); LBL-SLAC Collaboration LBL-3897, SLAC-PUB-1599 (1975) and private communication.
5. A. de Rujula and S. L. Glashow, Phys. Rev. Letters 34, 46 (1975); Thomas Appelquist et al., Phys. Rev. Letters 34, 6 365 (1975); E. Eichten et al., Phys. Rev. Letters 34, 6, 369 (1975).
6. E. Pelaquier and F. M. Renard, Preprint, Departement de Physique Mathematique U.S.T.L., 34060 Montpellier, Cedex, France.
7. F. A. Berend, R. Gastman, Nuclear Physics B61, 414 (1973).
8. J. Heintze, Proceedings of the 1975 International Symposium on lepton and photon interactions at high energies (Stanford 1975), p. 97.
9. J. W. Simpson et al., Phys. Rev. Letters 35, 699 (1975).

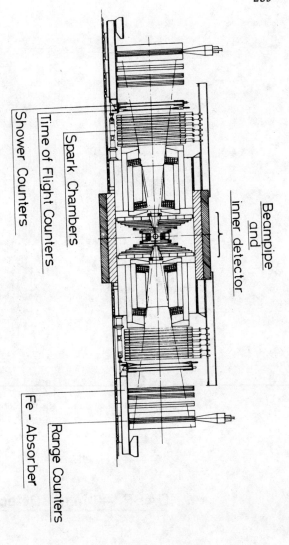

Fig. 1 DASP detector viewed along the beam direction.

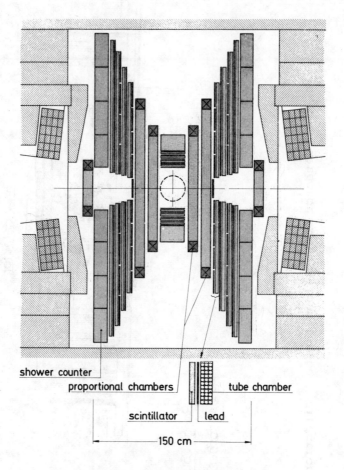

DASP — Inner Detector

Fig. 2 The inner detector of DASP viewed along the beam direction.

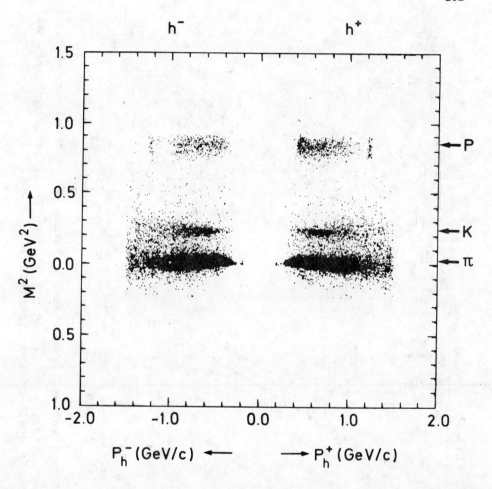

Fig. 3 The mass of the hadrons, as measured by time of flight, via their momentum, as determined with the spectrometer.

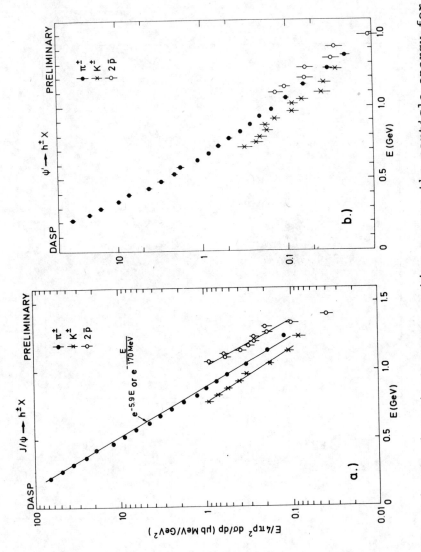

Fig. 4 The invariant cross section versus the particle energy for the decay a.) $J/\psi \to h^{\pm} X^{\mp}$ and b.) $\psi' \to h^{\pm} X^{\mp}$. The uncertainty in the absolute scale of the cross section is 30%.

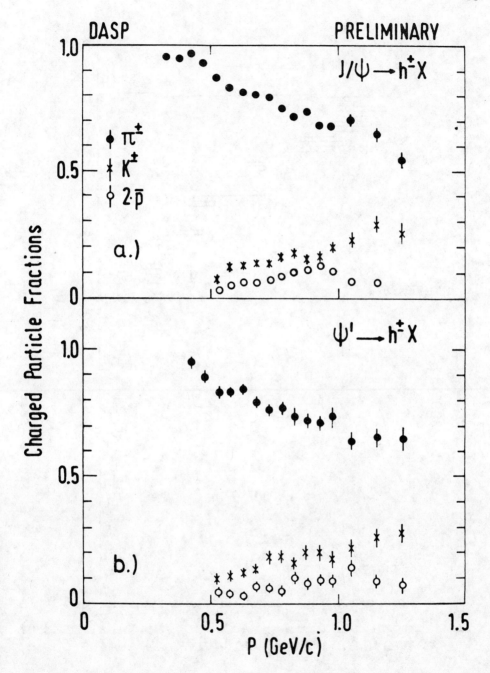

Fig. 5 Fractions of pions, kaons and antiprotons (x 2) observed a.) at the J/ψ resonance and b.) at the ψ' resonance.

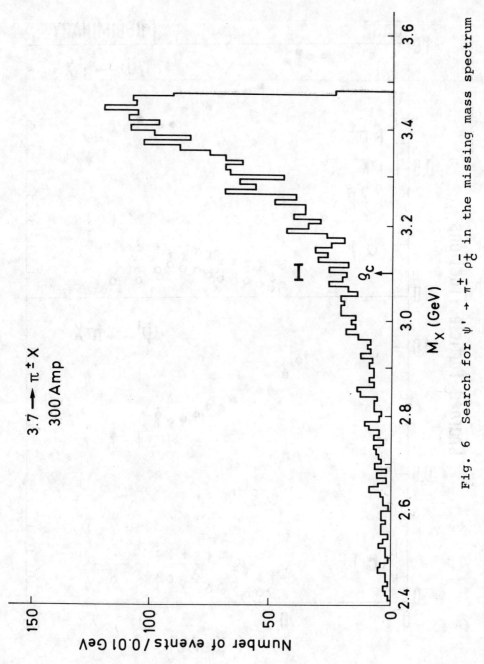

Fig. 6 Search for $\psi' \to \pi^{\pm} \rho^{\mp}_{\xi}$ in the missing mass spectrum

265

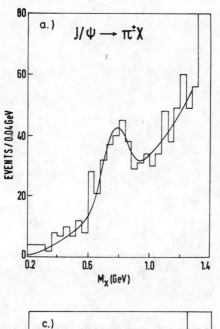
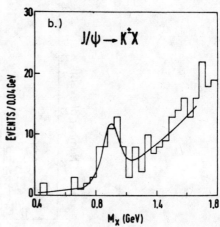
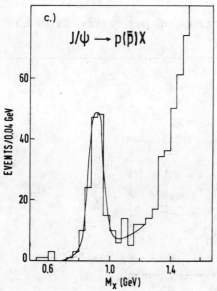

Fig. 7 Missing mass spectrum observed in the decay a.) $J/\psi \rightarrow \pi^{\pm} X^{\mp}$, b.) $J/\psi \rightarrow K^{\pm} X^{\mp}$ and c.) $J/\psi \rightarrow p X^-, \bar{p} X^+$. A ρ, K^* and proton antiproton peak is clearly seen in a.) b.) and c.) respectively.

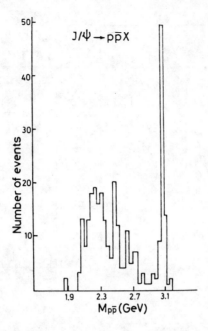

Fig. 8 Effective mass spectrum of p p̄ pairs from the decay of the J/ψ resonance.

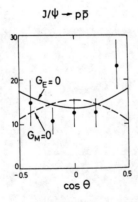

Fig. 9 The angular distribution of the events J/ψ → p p̄. The solid and dashed lines represent the fact that the electric and the magnetic nucleon form factor are respectively zero.

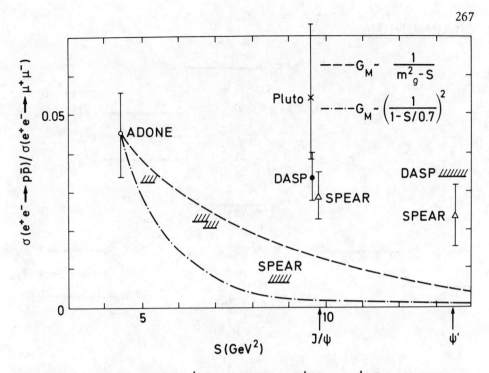

Fig. 10 The ratio $\sigma(e^+e^- \to p\bar{p})/\sigma(e^+e^- \to \mu^+\mu^-)$ as measured in this experiment (ϕ) and from Ref. (4). The upper limits (////) are obtained from measurements of $p\bar{p} \to e^+e^-$, $\mu^+\mu^-$ (Ref. 6) and measurements of $e^+e^- \to p\bar{p}$ (this experiment and LBL-SLAC Collaboration at SPEAR). The curves show a ρ-pole and a dipole behaviour of G_M normalized at the point $s = 4.4$ GeV^2.

CHARMONIUM

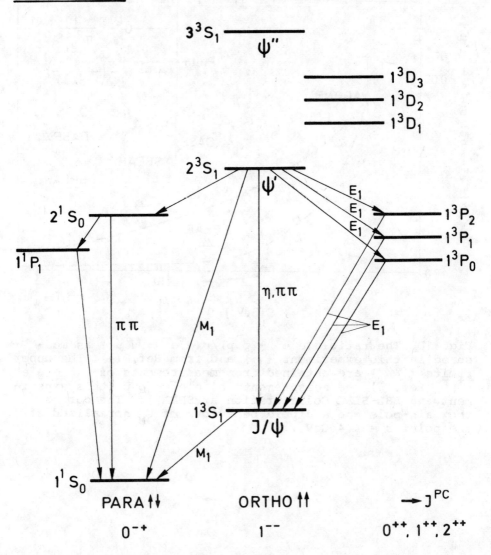

Fig. 11 Narrow states and their transitions as predicted by the charmonium model.

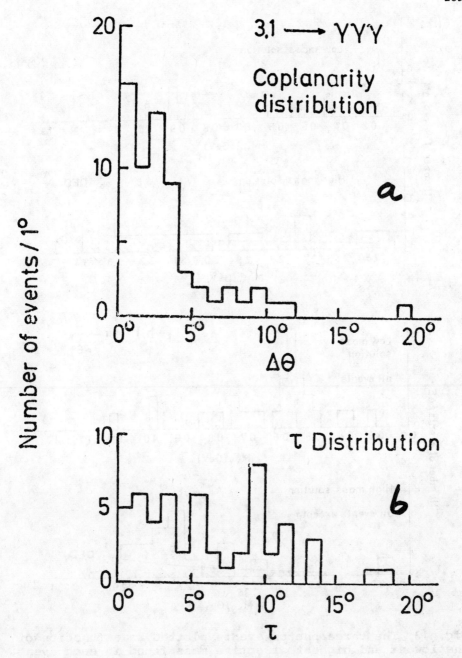

Fig. 12 a.) Number of three photon events versus the coplanarity angle $\Delta\theta$.

b.) Number of three photon events with $\Delta\theta < 5°$ versus the angle τ (defined in the text).

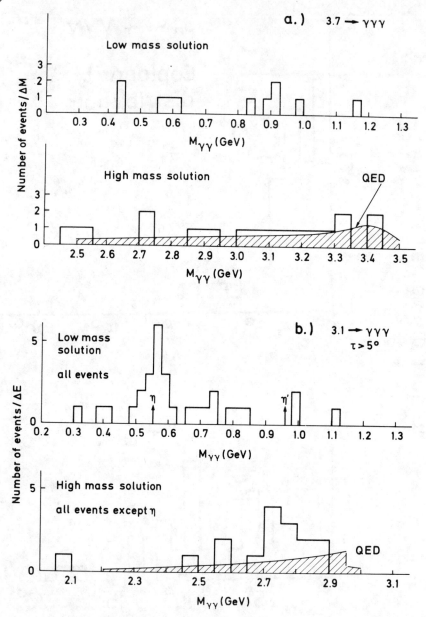

Fig. 13 The three photon events plotted as a function of the lowest and highest effective mass found in each event, as measured a.) at the ψ' resonance and b.) at the J/ψ resonance. Events with the low (or intermediate) effective mass within $\pm\sigma$ of the mass of an η or η' are removed from the plot.

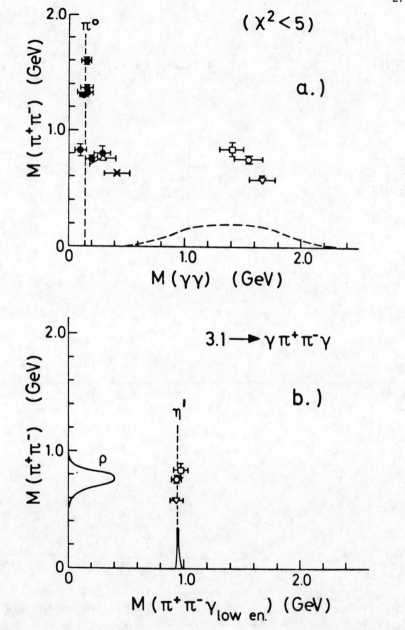

Fig. 14 a.) Scatter plot of $J/\psi \to \pi^+\pi^-\gamma\gamma$ events with no requirement on the opening angle of the two photons.
b.) Scatter plot of $J/\psi \to \pi^+\pi^-\gamma\gamma$ events after removing all events with an opening angle between the two photons smaller than $100°$.

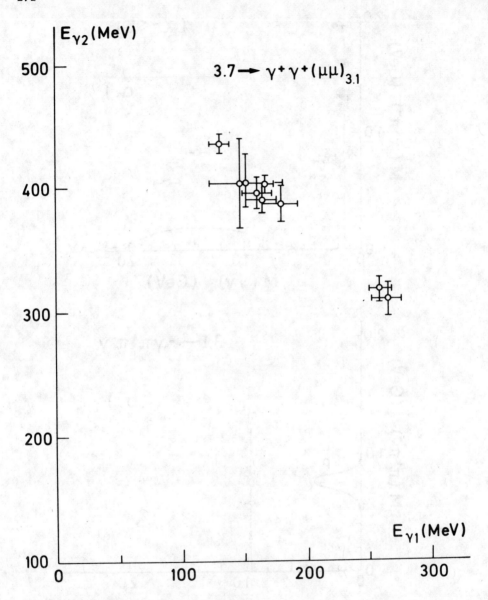

Fig. 15 Scatter plot of the two photon energies for the decay $\psi' \to (J/\psi \to \mu^+\mu^-) + \gamma\gamma$.

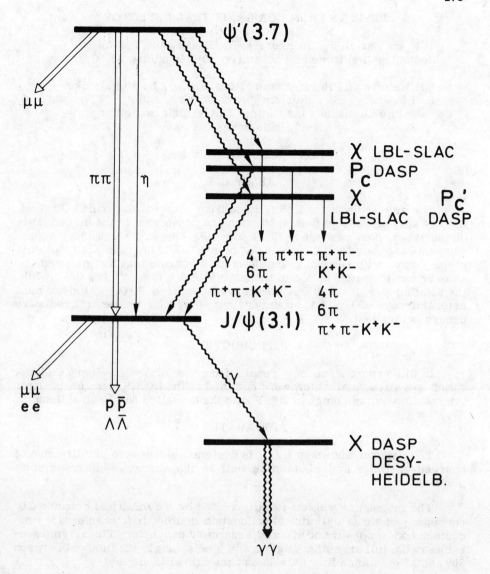

Fig. 16 The narrow states J/ψ, ψ', P_c, χ and X and their observed transitions.

RESULTS FROM DORIS: NEUTRAL DETECTOR[*]

W. Bartel, P. Duinker, J. E. Olsson, and P. Steffen
Deutsches Elektronen-Synchrotron, D-2 Hamburg 52, Germany

J. Heintze, G. Heinzelmann, R. D. Heuer, R. Mundhenke,
H. Rieseberg, B. Schürlein[**], A. Wagner, and A. H. Walenta
Physikalisches Institut der Universität Heidelberg,
D-69 Heidelberg, Germany

Presented by H. Rieseberg

ABSTRACT

The apparatus of the DESY-Heidelberg group is described. Preliminary results of a recent data evaluation are presented. In the investigation of the two photon cascade $\psi'(3.7) \to \gamma + P_C$, $P_C \to \gamma + J/\psi (3.1)$ about 60 events are found with $J/\psi \to \mu^+\mu^-$ or e^+e^-. The angular distributions of the γ-rays with respect to the beam axis indicate that the intermediate state P_C has non-zero spin and the mass $M(P_C) = (3.51 \pm 0.02)$ GeV, while the solution $M(P_C) = 3.26$ GeV seems to be excluded. From a study of total neutral decays of $J/\psi (3.1)$ an upper limit of 1.2% for the neutral radiative decays is obtained.

INTRODUCTION

In this report some new results from the DESY-Heidelberg Collaboration are presented which were obtained at the DORIS electron-positron intersecting storage rings at DESY with the so-called NaI Neutral Detector.

APPARATUS

The detector shown in fig. 1 is designed to measure the direction of charged particles and photons as well as the energy of electrons and photons.

The interaction region is surrounded by a cylindrical detector CD consisting of three cylindrical concentric double drift chambers[2], two counter hodoscopes H and H', and a mercury converter. The arrangement encloses the full azimuthal angle φ and a polar angle (ϑ) interval between 30° and 150°, subtending thus 86 % of the full solid angle.

Each double drift chamber comprises 128 anode wires for the measurement of the ψ co-ordinate and 80 cathode strips for the ϑ co-ordinate. To facilitate computations, the wires and the strips of the three chambers are aligned radially to the center. The accuracy obtained for charged tracks in the angle measurement is $\Delta\varphi \approx 4$ mrad, $\Delta\vartheta \approx 20$ mrad at $\vartheta = 30°$

---

[*] Work supported by the Bundesministerium für Forschung und Technologie
[**] Now at IDAS GmbH, D-625 Limburg

and $\Delta\vartheta \approx 40$ mrad (fwhm) at $\vartheta = 90°$. The timings of both the anodes and the cathodes are measured. As the induced cathode signal has the same timing as the primary signal of the anode wire, $\varphi-\vartheta$ ambiquities can be mostly removed in multiprong events by comparing drift times.

The scintillation hodoscopes H and H' are composed of 32 elements each. An element covers an azimuthal interval $\Delta\varphi = 22.5°$ and a polar angle interval $\vartheta = 30°$ to $90°$ or $90°$ to $150°$. A track is identified by a coincidence between corresponding cells of the internal and external hodoscopes.

The region of small scattering angles ($15° \leq \vartheta \leq 30°$ and $150° \leq \vartheta \leq 165°$) is covered by two additional scintillation counter hodoscopes M, of 16 elements each. By these hodoscopes, which are not included in the

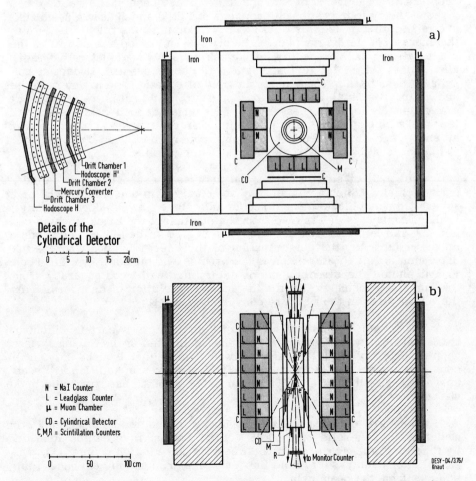

Fig. 1. Apparatus of the DESY-Heidelberg Experiment: a) Elevation section viewed along the beam axis and enlarged sector of the cylindrical detector, b) plan section.

trigger, the solid angle for the detection of charged particles is enlarged to 95 % of the full solid angle.

A vessel made of epoxy-fibre glass is positioned between the second and the third chamber. It can be filled with mercury. Photons originating from the interaction point see a converter thickness of two radiation lengths. The co-ordinates of the conversion points are measured in the third chamber with accuracies $\Delta\varphi \approx 35$ mrad (fwhm) and $\Delta\vartheta \approx 65$ mrad (table I).

The cylindrical detector is surrounded by an arrangement of NaI (N) and lead glass (L) counter blocks which allow to identify electrons and photons and to measure their energies in a solid angle of 60 % of 4π. The NaI counters are also used to measure the energy loss, and the lead glass counters to measure β^2 of hadrons and muons. Each of the walls on both sides of the cylindrical detector consists of 12 NaI and 20 lead glass blocks, and is 13 radiation lengths thick. The NaI counters improve considerably the energy resolution for photons below 500 MeV. Top and bottom of the inner detector are covered by lead glass bars (18 centimetres $\triangleq$ 8 radiation length thick, 140 centimetres long). Phototubes are attached to both ends of these bars to compensate for light attenuation and to determine the shower position by a time difference measurement. All pulse heights are mixed linearly to get a measure of the energy deposited in the NaI and lead glass blocks. The sum signal is used in the trigger. The response of all energy measuring detectors is continuously monitored by observation of Bhabha-scattered electrons and of passing muons.

A deposit of energy in the NaI and leadglass detector is called an energy cluster. Such a cluster may be contained in one single block or in several adjacent blocks. The direction of a photon, which does not convert in the mercury, is determined from the position of the blocks, which contain the shower, and from the ratio of energies in the blocks. For such a unconverted γ-ray the determination of the energy is good whereas the determination of the direction is not precise. To the contrary, for a converted photon the direction can be determined with good accuracy from the signals in chamber 3 while the energy resolution is deteriorated by the energy loss in the mercury converter (table I).

The trigger consists of several combinations of a certain minimum track multiplicity with a minimum energy deposited in the NaI-lead glass counter (see table II). The completely neutral trigger is used only, if the converter vessel is filled with mercury. It requires in addition to a seen energy of more than 1500 MeV that at least one photon has converted in the mercury.

Muons are detected in drift chambers (μ) behind a 60 cm thick iron moderator over a solid angle of 45 % of 4π. Cosmic rays can be identified by time of flight measurement between the scintillation counters C. The scintillation counters R around the beam pipe are used to veto background triggers from the beam halo.

Table I. γ-rays: Precision of conversion point and energy

Place of conversion	Method to determine the conversion point	Δφ (fwhm) (mrad)	Δϑ (fwhm) (mrad)	Resolution of the energy measurement
NaI and lead glass blocks	energy cluster in Side Wall	240	100	good
	energy cluster in Top and Bottom	150	240	
Hg converter	centre of the hit wires in chamber 3	75	65	distorted by loss in converter
	centre of the shower determined from the electron tracks in chamber 3	35	65	

Table II. Trigger conditions

Number of prongs seen in HH'	Minimum energy recorded (MeV)
≥ 3	300
2	400
1	900
0	1500

A PDP 9 on-line computer selects in a fast decision only such events which contain tracks originating from the zone around the interaction point. For the zero-prong trigger at least one shower in chamber 3 and an energy cluster in the NaI/lead glass blocks behind the conversion are required.

Yields are normalized by measurements of Bhabha scattering at small ϑ angles between 105 and 155 mrad.

The DORIS storage rings produce an interaction region which is about 20 mm (fwhm) long and in the plane transverse to the beam 0.6 mm (fwhm) high and 0.4 mm (fwhm) wide. The energy distribution of electrons within a beam bunch is approximately Gaussian with a rms width of about 0.5 MeV at $E_{c.m.}$ = 3.1 GeV.

PROPERTIES OF THE INTERMEDIATE STATE P_C IN THE ψ' (3.7) RADIATIVE DECAY CHAIN

In July 1975, the DASP-collaboration at DESY discovered[3] an intermediate state, P_C, in the photon decay sequence:

$$\psi'(3.7) \rightarrow \gamma\gamma\, J/\psi(3.1) \tag{1}$$

This discovery was confirmed by our group[1] and by the SLAC-LBL

Collaboration[4] leaving open the question if P_C is located near 3.50 GeV or near 3.27 GeV.

We present here preliminary results on a re-examination of the P_C intermediate state. The aim of this research was to study the properties of P_C in a sample of cascade events where $J/\psi(3.1)$ is identified via its leptonic decays:

$$\psi'(3.7) \to \gamma\, P_C, \quad P_C \to \gamma\, J/\psi(3.1) \tag{2}$$

$$J/\psi(3.1) \to \begin{Bmatrix} \mu^+\mu^- \\ e^+e^- \end{Bmatrix} \quad \begin{matrix}(3)\\(4)\end{matrix}$$

We first deal with events where the $\psi(3.1)$ decays into a muon pair.

Starting from a condensation of events with two prongs and ≥ 1 gammas the following selection criteria were imposed on the events by a computer program and a succeeding visual inspection:

1. Only two charged tracks have to be seen in the central detector and at least one of these has to give a signal in a muon chamber. This criterion removes charged hadrons and electrons from the sample.

2. Two photons are required to deposit energy in the NaI and lead glass detector. The energy of each cluster has to be greater than 30 MeV, and the blocks, which contain the cluster, have to be more than 100 mrad in angle away from the muon tracks. This criterion asks for two detected photons which are well separated from the two tracks. A conversion in the mercury converter is not required, but for events without conversion the total recorded γ-energy has to exceed 500 MeV.

3. To exclude events with additional tracks and photons, those events were removed from the sample which have the small angle hodoscope M fired or contain converted photons whose energy is not recorded.

4. Events which are generated by cosmic showers are also eliminated.

131 events remain after these procedures.

The kinematics of the decay chain under discussion (reactions 1 and 3) is completely determined by the direction of the four final state particles. The measured γ-energies constrain the event and allow a fit. If both γ-ray showers are fully contained in the NaI and lead glass blocks the fit has two constraints. In case a shower hits the edges of the blocks, the energy measurement is not used as a constraint.

In the following the further steps towards the final sample are shortly described.

In a first step, a fit to the hypothesis

$$\psi'(3.7) \to \gamma\gamma\,\mu^+\mu^- \tag{5}$$

is tried. The fit uses the decay vertex (calculated from the charged tracks), the direction of the charged particles, the direction of the γ-rays, and the γ-energies (if available) as measured quantities. A fit is declared to be successful if the probability for the hypothesis is bigger than 0.01.

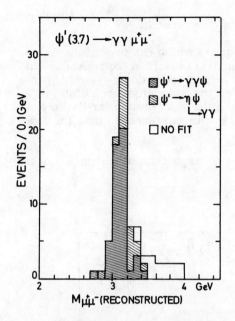

Fig. 2. Invariant mass distribution of $\mu^+\mu^-$ pairs from the decay $\psi'(3.7) \to \gamma\gamma\mu^+\mu^-$. If the $\mu^+\mu^-$ mass is constrained to the mass of the $J/\psi(3.1)$ 18 events give a successful fit under the $\psi' \to \eta\psi$, $\eta \to \gamma\gamma$ hypothesis and 49 events under the $\psi' \to \gamma\gamma J/\psi$ hypothesis. 14 events give no fit.

81 events pass the fit. The distribution of the invariant mass of the muon pair is shown in fig. 2. It clusters around 3.1 GeV.

In a second step, a fit to the same hypothesis (5) is made, but the muon pair mass is constrained to 3.1 GeV. 14 events give no fit. They are probably due to the decay

$$\psi'(3.7) \to \pi^0 \pi^0 J/\psi(3.1), \quad (6)$$
$$\downarrow \gamma\gamma$$
$$\downarrow \gamma\gamma$$

where two of the four γ-rays escape detection through the gaps of the detector. 18 events fit under the hypothesis of the known decay mode [5]

$$\psi'(3.1) \to \eta J/\psi(3.1). \quad (7)$$
$$\downarrow \gamma\gamma$$

The $\gamma\gamma$ angular distribution and the $\gamma\gamma$ mass of these events correspond exactly to the expectation for this decay.

We are left with 49 events which agree with hypothesis (1). A Monte Carlo simulation shows that they contain a background of 10 % from events of reaction (6). The reconstructed γ-energies are plotted in fig. 3a. There are two correlated sharp peaks at 167 MeV and 398 MeV, which give a clear evidence for the intermediate state P_C. The energy resolution is not sufficient to distinguish between the first and the second gamma by Doppler broadening.

To increase statistics, we now add events[1] of type 4 where the $J/\psi(3.1)$ decays into e^+e^-. The electron data were re-analysed with criteria adjusted to electrons (e.g. the electron energies serve as input to the fit). The combined γ-ray spectrum (77 events) is shown in fig. 3b together with the calculated background ($\approx$ 8 events) from reaction (6).

To investigate the spin properties of the P_C state the highest energy γ is confined to the interval 350 to 450 MeV, leaving 60 events of reactions (2) to (4) with a background of 3 events of reaction (6).

By the production through an e^+e^- collision, the spin of the $\psi'(3.7)$ is aligned with the beam direction. (The beams in DORIS are not polarized.) We call the angle between the beam axis and one of the gammas Θ. The angular distribution as a function of $\cos\Theta$ is plotted in fig. 4. We observe a flat distribution for the high energy γ, whilst the distribution of the low energy γ is strongly peaked at 90°.

In a γγ cascade, no mechanism exists, which can produce a $\sin^2 \Theta$ unisotropy for the second γ-ray, if the first is emitted isotropically. In order to obtain an unisotropy for the second transition $P_C \to \gamma J/\psi$, the first transition $\psi' \to \gamma P_C$ has to transfer the spin alignment of the ψ'. This is only possible with a non-flat angular distribution for the first γ. Therefore, it is very probable, even within our errors, that the low energy γ is the first one.

As the $\psi'(3.7)$ has spin 1, one has to expect for the emission of the first γ a distribution

$$W(\Theta) = 1 + \cos^2 \Theta, \tag{8}$$

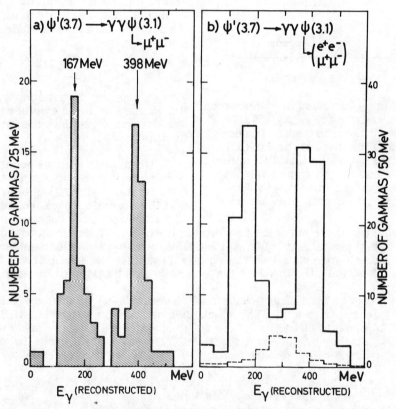

Fig. 3. Photon energy spectrum for candidates for the decay chain $\psi'(3.7) \to \gamma\gamma J/\psi(3.1)$, two entries per event. a) $J/\psi(3.1) \to \mu^+\mu^-$ (49 events), b) in double-wide binning the same events as (a) combined with 28 events where $J/\psi(3.1) \to e^+e^-$. The dashed histogram in (b) represents the calculated background from the reaction $\psi'(3.7) \to \pi^0\pi^0 J/\psi(3.1)$.

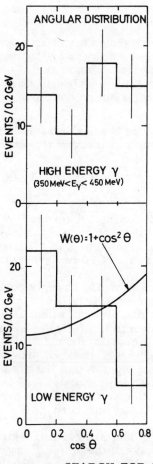

if for the P_C spin zero is assumed. The observed distribution of the low energy γ is in contradiction to this expectation. It is not possible to distinguish between spins 1 and 2 in this stage of the analysis[6].

We conclude therefore: The intermediate state P_C has non-zero spin and its mass is

$$M(P_C) = (3.51 \pm 0.02) \text{ GeV}. \quad (9)$$

A more detailed analysis is in progress.

Fig. 4. Angular distribution of the high energy and of the low energy γ-ray in the decay chain $\psi' \to \gamma P_C$, $P_C \to \gamma J/\psi$. Θ is the angle between the beam direction and the direction of the γ-ray. The energy of the high energy γ is confined to the interval between 350 and 450 MeV.

SEARCH FOR NEUTRAL DECAYS OF THE ψ(3.1)

The decay of the $J/\psi(3.1)$ into final states containing at least two charged particles has been extensively studied by the SLAC-LBL[7,8] and DASP[9] collaborations. The Gamma-Gamma Group at ADONE has measured the photon multiplicity distributions down to only one detected charged prong[10].

We report here an analysis of final states containing more than three γ-rays and no charged particles.

The events were selected by the following criteria from the zero-prong trigger (energy threshold 1500 MeV).

1. The total recorded energy in one quadrant of the apparatus (one side wall, top or bottom detector) has to exceed 650 MeV.
2. At least three γ-clusters with individual energies above 8 MeV have to be found in the apparatus.

3. Two conversions in the mercury are required, one with $\vartheta < 90°$ and one with $\vartheta > 90°$.
4. The angle $\Delta\varphi$ between any two (in φ) adjacent γ-clusters of an event has to be $\leq 180°$. This means that the measured energy is not concentrated into one half of the apparatus (half apparatus condition). This condition eliminates cosmic showers which hit the NaI/lead glass blocks, the external hodoscope H and chamber 3 but do not cross the central part of the detector.
5. Events with a hit in the small angle hodoscope M are also excluded, since one does not know, if such a hit is due to a track or to a converted γ-ray.

After the selection, 2624 events remain from about 740 000 observed J/ψ decays.

A hand scan of a limited sample gives the following background: 6 % cosmic ray events, 4 % events with tracks passing through gaps of the internal hodoscope, and 4 % misidentified QED two γ-events.

From the scan we obtain the experimental γ-multiplicity distribution which is shown in fig. 5. These events still contain about 1.5 % of events with charged prongs escaping through the beam pipe (solid angle: 5 % of 4π). Photon multiplicities as high as 11 are observed.

The observed all-neutral decays of the $J/\psi(3.1)$ have two possible sources: Radiative decays and hadronic decays.

We consider first the radiative decays. The J/ψ is a C = -1 state, while N $\pi°$ or N η are C = +1 states. Therefore, if C-conservation is assumed, the decay of the J/ψ into neutral pions or etas is expected to proceed via a radiative decay of the form:

$$J/\psi(3.1) \rightarrow \gamma + N\pi° \quad (10)$$

or $J/\psi(3.1) \rightarrow \gamma + N\eta \quad (11)$

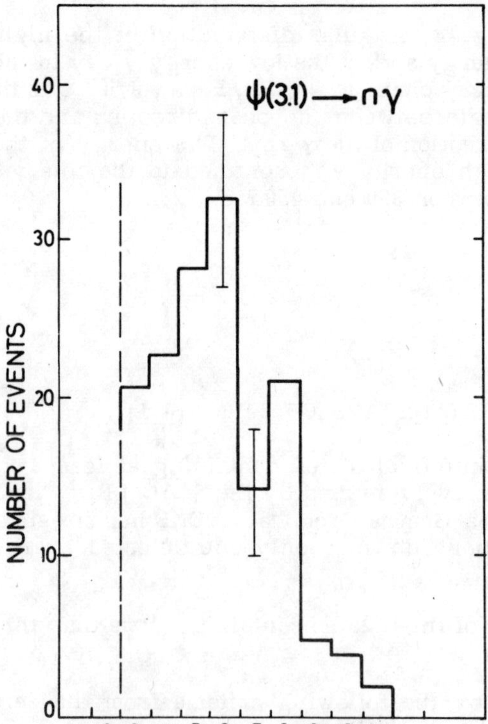

Fig. 5. Experimental multiplicity distribution of γ-rays in neutral decays of the $J/\psi(3.1)$ resonance. Shown here is a sample of 151 events from a total of 2250 events.

We have calculated by a Monte Carlo simulation the

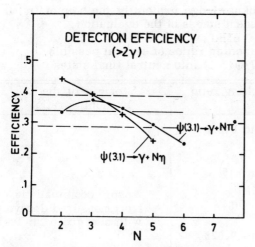

Fig. 6. Result of a Monte Carlo simulation: Efficiency ε for detecting more than two gammas from radiative decays $J/\psi(3.1) \to \gamma + N\pi^0$ and $J/\psi(3.1) \to \gamma + N\eta$, plotted versus the number N of all neutral pions or etas. The average efficiency $\langle \varepsilon \rangle$ = 0.33 ± 0.05 in the range $2 \leq N \leq 5$ is indicated.

efficiency ε to detect such radiative decays assuming an invariant phase space model for reactions (10) and (11). As can be seen from fig. 6, the efficiency varies only slightly with the total number N of neutral pions or etas. The variation can be qualitatively unterstood: While the cut-off at small N comes from the selection criteria 2 to 3, the decrease at high N is due to the increasing number of beam pipe conversions. For the interval $2 \leq N \leq 5$ the average detection efficiency is

$$\langle \varepsilon \rangle = 0.33 \pm 0.05. \qquad (12)$$

We obtain from the observed event number (2250 after background subtraction), from the total number of seen J/ψ decays (740 000) and from $\langle \varepsilon \rangle$ the following upper limit for the neutral radiative decays of the $J/\psi(3.1)$:

$$\frac{\Gamma(J/\psi \to \gamma + N\pi^0 + M\eta)}{\Gamma(J/\psi \to \text{all})} < 1.2 \% \qquad (13)$$

Only an upper limit is given since the fraction due to hadronic decays into neutral final states is not known. A rough estimate of this constribution from experimental known decays (table III) however explains at most one fifth of the data, leaving the rest for radiative decays or yet unknown hadronic decay modes.

We emphasize that the 3 γ decays were excluded from this consideration. Preliminary results from the $J/\psi \to 3\gamma$ channel were presented at the Stanford conference[1]. There was a clear η signal and some indication for a η' signal and a new state (X) near 2.8 GeV. An evaluation of all data, taken both before and after the Stanford conference, and a detailed study of the background, induced by higher γ multiplicities in the 3 γ sample, are in progress.

ACKNOWLEDGEMENT

We wish to express our thanks to the members of the DORIS machine group under the leadership of Dr. D. Degèle for the effective operation of the storage rings, and to the members of the DESY S2 group for their important help during the installation of the experimental equipment. Particular thanks are due to our engineer H. Matsumura and to our technicians J.-H. Seidel and J. Zimmer for their assistance in various stages of the experiment. We acknowledge gratefully the help of Dr. D. Pandoulas who participated in recent stages of the evaluation.

Table III. Calculated branching ratios of several possible hadronic decays of the $J/\psi(3.1)$ into neutral final states

Decay	branching ratio (%)	origin of the branching ratio
$2\pi^0 K_S^0 K_L^0$ $\hookrightarrow 2\pi^0$	0 to 0.062	
$K^0 K_{892}^{0*} \to 3\pi^0 K_L^0$	0.012	isospin combinations from measured decays with charged particles[8]
$K_{892}^{0*} K_{1420}^{0*} \to 4(5)\pi^0 K_L^0$	0.004	
$4\pi^0 K_S^0 K_L^0$ $\hookrightarrow 2\pi^0$	0 to 0.023	
$2\pi^0 \omega$ $\hookrightarrow \pi^0 \gamma$	0.035	
$\Lambda\bar{\Lambda} \to 2\pi^0 n\bar{n}$	0.020	measured[8]
$\Lambda\bar{\Lambda} \pi^0$	0.02	assumed: $\Gamma(\Lambda\bar{\Lambda}\pi^0) = \Gamma(\Lambda\bar{\Lambda})$
$n\bar{n} \pi^0$	0.03[11]	measured is $p\bar{p}\pi^0$ etc.[8]
	minimum sum: 0.12 maximum sum: 0.21	

REFERENCES AND FOOTNOTES

1. J. Heintze, Proceedings of the 1975 International Symposium on Lepton and Photon Interactions at High Energies, Stanford (1975), p. 97.
2. J. Heintze and A. H. Walenta, Nucl. Instr. Meth. 111, 461 (1973); A. H. Walenta, IEEC Transactions on Nuclear Science 22, 251 (1975); A. H. Walenta, RHEL/M/H21, 116 (1972)
3. DASP Collaboration, Phys. Lett. 57 B, 407 (1975)
4. W. Tanenbaum et al., Phys. Lett. 35, 1323 (1975)
5. W. Tanenbaum et al., Phys. Lett. 36, 402 (1976)
6. P. K. Kabir and A. J. G. Hey, Preprint (1975)
7. B. Jean-Marie et al., Phys. Rev. Lett. 36, 291 (1976)
8. G. S. Abrams, Proceedings of the 1975 International Symposium on Lepton and Photon Interactions at High Energies, Stanford (1975), p. 25
9. B. H. Wiik, Proceedings of the 1975 International Symposium on Lepton and Photon Interactions at High Energies, Stanford (1975), p. 69
10. R. Baldini-Celio et al., Phys. Lett. 58 B, 471 (1975)
11. This branching ratio is reduced by a factor 4 because of the expected lower detection efficiency for a decay with only one neutral pion. No such reduction is included in the other figures.

INCLUSIVE PARTICLE PRODUCTION AND ANOMALOUS MUONS
IN e^+e^- COLLISIONS AT SPEAR*

Bruce A. Barnett

University of Maryland, College Park, Maryland 20740
and
The Johns Hopkins University, Baltimore, Maryland 21218

ABSTRACT

Results are presented on hadron and muon inclusive production in e^+e^- collisions at $\sqrt{s}$ = 3.8 and 4.8 GeV. Anomalously large high momentum muon production is observed in noncoplanar two charged particle final states, but no anomalies are seen in multicharged particle final states. Arguments are presented that these extra muons do not come from charmed particles but could possibly come from heavy leptons. Results are also presented on a search for e^+e^- narrow resonances with 5.7 GeV $< M_{ee} <$ 6.1 GeV. The $T_{(5.97)}$ was not seen in this scan which means that $\Gamma_{ee} <$ 100 ev. if the T has decay modes similar to the Ψ and Ψ'.

INTRODUCTION

I would like to present some of the results from two different e^+e^- experiments which have been performed at the SPEAR facility at SLAC. The two experiments are SP8, a Maryland, Pavia, Princeton collaboration, and SP14/19, a Maryland, Pavia, Princeton, U.C. San Diego, SLAC collaboration. The physicists associated with these experiments are as follows:

SP8

T. L. Atwood[**,#], D. Badtke[**], B. S. Barnett[**,δ], M. Cavalli-Sforza[+], D. G. Coyne[#], G. Goggi[+], G. C. Mantovani[+], G. K. O'Neill[#], A. Piazzoli[+], B. Rossini[+], H. F. W. Sadrozinski[**,φ], D. Scannichio[+], K. Shinsky[#], L. V. Trasatti[**,φ], G. T. Zorn

SP 14/19

D. Aschman[#], D. Badtke[**][+], B. A. Barnett[**][δ][#], C. Biddick[α],
T. Burnett[α], M. Cavalli-Sforza[+][**], D. G. Coyne[β], G. Goggi[+], D. Groom[β][#],
F. Impellizeri[φ], L. Jones[α], L. Keller[β], M. Liyan[α], D. Lyon[+], G.
Masek[α], E. Miller[α], G. K. O'Neill[#], F. Pastore[α], B. Rossini[β], H. F. W.
Sadrozinski[#], K. Shinsky[**], J. Smith[α], J. Stronski[+], M. Sullivan[α],
W. Vernon[α], G. T. Zorn.

The procedure I will follow in this talk will be to discuss the SP8 results first and then turn to the SP14/19 results. The high momentum hadron spectra from SP8 have been published[1] for some time, so I will review that data only to the extent that it bears on my main topic, which is the inclusive muon spectrum. This spectrum shows an anomalously large number of muons[2] in final states having two charged particles with large noncoplanarities in comparison to what is expected from QED or hadron effects. These extra muons could be related to some new process like charm, heavy lepton, or quark-gluon production. The SP14/19 results which I'll discuss comes from a recent energy scan in the 5.7 → 6.1 GeV region, where we looked for narrow resonances in the total cross section. Preliminary analysis of the data shows no such resonances which puts interesting limits on the coupling of $T_{(5.97)}^3$ to e^+e^-.

---

[*]Work supported by the Energy Research and Development Administration, Istituto Nazionale di Fisica Nucleare, and the National Science Foundation.

[**]University of Maryland, College Park, Maryland

[+] Istituto di Fisica Nucleare, Universita di Pavia and Istituto Nazionale di Fisica Nucleare, Sezione di Pavia, 27100 Pavia, Italy

[#]Princeton University, Princeton, New Jersey

[α]U. C. San Diego, San Diego, California

[β]Stanford Linear Accelerator Center, Stanford, California

[φ]Present address: Laboratori Nazionali di Frascati, Casella Postale 70, Frascati, Rome, Italy

[δ]Present address: The Johns Hopkins University, Baltimore, Maryland

SP8 EXPERIMENTAL APPARATUS

The main characteristics of the SP8 apparatus are as follows:
1) The event trigger required only a single charged particle passing through the magnetic spectrometer.
2) The spectrometer was at $90°$ to the e^+e^- beams and covered about 1% of the total solid angle.
3) The experiment made $e/\mu/\pi/K/p$ identification for particles of any momentum passing through the spectrometer by means of a Cerenkov counter, shower counters, range-hadron filter, and time of flight.
4) Particles traveling away from the spectrometer on the opposite side of the e^+e^- beams were partially identified by a shower counter and range-hadron filter.
5) A central detector, called the "polymeter", covered 99% of the 4π steradian solid angle and measured the charge multiplicity of events which triggered our system. It consisted of four units (one above, below, and on each side of the beam), each containing three proportional wire chambers having wires parallel to the e^+e^- beams. It measured the ϕ angle of all charged particles, but made no θ measurement.

The Figure 1 shows a schematic of the apparatus as seen from above. The magnetic field was vertical and rather uniform at $\approx$ 4.2 K gauss; the total $\int Bd\ell$ was $\approx$ 11.8 K gauss-meters. The positions of all of the proportional wire chambers are shown, where the label of "X" means the wires run vertically and "Y" means the wires run horizontally. The time of flight measurement was made between the scintillation counter, S1, and the first two scintillation counter planes of the shower counter to give a total flight path of about 5 meters. The Cerenkov counter was filled with 90 psig of propane giving a π threshold of 1.05 GeV/c. The shower counters were made of a five layer sandwich of lead plates and scintillation counters. The total thickness of the counters were 7.2 radiation lengths and the average electron pulse was $\approx$ 6 times that of a minimum ionizing muon. The hadron filters consisted of 69 cm of iron with three interspersed planes of five scintillation counters placed side by side. The momentum needed by a muon to penetrate the iron was $\approx$ 1.05 GeV/c, the exact value depending on the angle of incidence, scattering, and straggling[4]. The conjugate side shower counter and hadron filter, covering respectively 2.5 sr. and 1.7 sr., identified back to back electrons and muons.

HADRON SPECTRA ANALYSIS

The description of the apparatus should make it clear that the general definition of a hadronic event requires a non-showering, non-penetrating particle in the spectrometer. The normalization of the experiment was achieved through the number of $e^+e^- \to \mu^+\mu^-$ events which can be related to the integrated luminosity by calculations assuming the validity of QED. This should be a good assumption since several experiments[5] have showed that $e^+e^- \to \mu^+\mu^-$ and

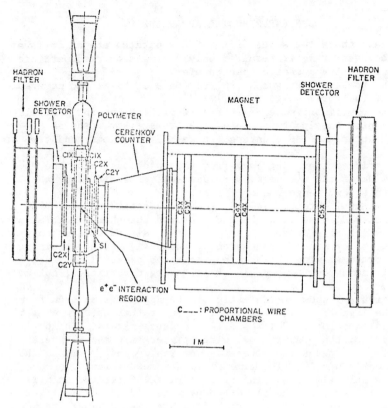

Fig. 1

$e^+e^- \to e^+e^-$ events with small noncollinearity agree with quantum electrodynamics. Figure 2 shows our noncollinearity distribution for $\mu^+\mu^-$ events defined by having penetrating particles in both hadron filters at $\sqrt{s}$ = 4.8 GeV out to 30° which is the geometrical limit for total acceptance imposed by the conjugate side. The line drawn through the data is the result from a computer program of Berends, Gaemers, and Gastmans[6] which calculates $e^+e^- \to \mu^+\mu^-$ from QED to order α^3 using our experimental geometry. The agreement is very good, having χ^2 = 7.7 for 9 degrees of freedom. However, we limited our $\mu^+\mu^-$ data to 10° noncollinearity for normalization purposes since this corresponded to the cuts in many of the regular QED experiments. The QED calculations gave inclusive cross sections for $e^+e^- \to \mu^+\mu^-$ of $d\sigma/d\Omega$ = .44 nb/sr. at $\sqrt{s}$ = 4.8 GeV and $d\sigma/d\Omega$ = .69 nb/sr. at $\sqrt{s}$ = 3.8 GeV. Any cross sections quoted later have a normalization uncertainty of $\pm$ 7% at $\sqrt{s}$ = 4.8 GeV and $\pm$ 11% at $\sqrt{s}$ = 3.8 GeV due to the $\mu^+\mu^-$ statistics.

Now let me describe the hadron results. Corrections from our Monte Carlo program were applied to our raw hadron data to find the original production spectra and ratios. This gave a K/π ratio of 0.27 ± 0.08 and a p/π ratio of 0.04 ± 0.02 at $\sqrt{s}$ = 4.8 GeV for charged particles produced at 90° with momenta greater than 1.1

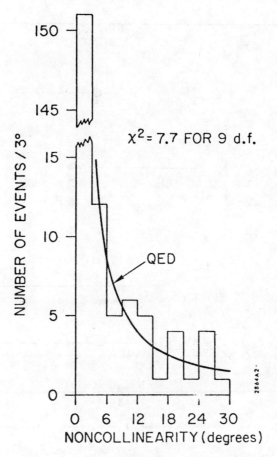

Fig. 2

GeV/c. The corresponding particle fractions are shown in Figure 3 along with the SLAC-LBL data[7] at lower momenta. Figure 4 shows the inclusive invariant cross section at 90°, differential in momentum for π's and K's at $\sqrt{s}$ = 4.8 GeV. This figure illustrates that both spectra fall off rapidly with increasing momentum, and that our statistics in the K/π ratio quoted above are dominated by the momentum region between 1.1 and 1.6 GeV/c where the pion cross section is distinctly larger than that of the kaons.

The charged multiplicity associated with the spectrometer hadron events was measured by the large solid angle "polymeter" surrounding the vacuum pipe near the interaction region. Figures 5, 6, and 7, show that quantity in various different forms: (5) the probability of seeing each multiplicity regardless of the spectrometer momentum or identity, (6) the event scatter plot of multiplicity versus momentum for each particle type, and (7) the average charge multiplicity versus momentum at $\sqrt{s}$ = 4.8 GeV. The few events with odd multiplicity are due predominately to γ-ray conversion in the 4%

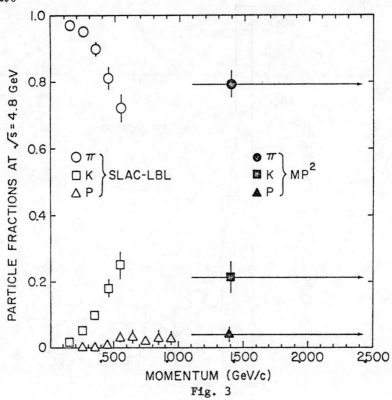

Fig. 3

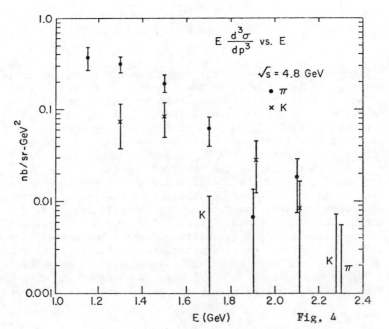

Fig. 4

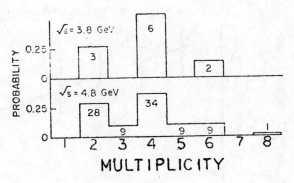

Fig. 5

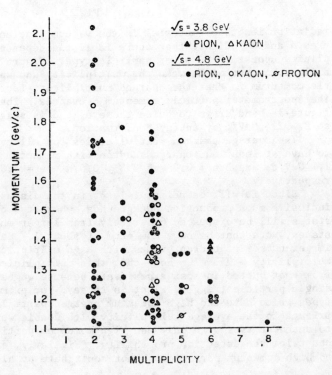

Fig. 6

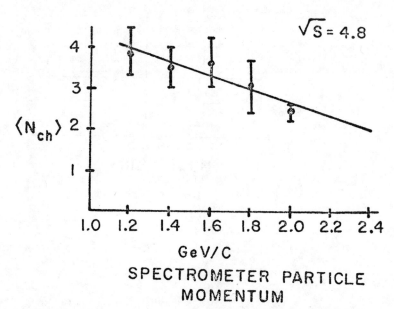

Fig. 7

radiation length of material in the vacuum pipe and first PWC. Figures 5 and 6 indicate that there is little dependence in the multiplicity upon either "s" or particle type. Figure 7, however, shows a clear connection between the multiplicity and spectrometer particle momentum in that the charged multiplicity increases rapidly as the spectrometer particle momentum decreases. The line in the figure is hand drawn to guide the eye, and is pegged to $<N> = 2$ at $E = 2.4$ GeV/c by energy conservation.

The average charged multiplicities for all events, selected to have at least one charged hadron with momentum greater than 1.1 GeV/c, are 3.6 ± 0.3 and 3.8 ± 0.5 at $\sqrt{s} = 4.8$ and 3.8 GeV, respectively.

Since it will be relevant later in the discussion of the inclusive muon production backgrounds, where two charged particle states will be discussed separately from larger multiplicity states, note that below a momentum of about 1.7 GeV/c there is a larger number of hadron events with high multiplicity than with multiplicity = 2. Also note that this multiplicity is different from that quoted in most experiments because we have an inclusive single particle trigger. That is to say, for example, that event types which have two high momentum charged particles will contribute to the average multiplicity with double weighting relative to an event type with only one high momentum particle because it has twice the detection probability. Similarly, event types with no high momentum particles do not contribute at all.

INCLUSIVE MUON PRODUCTION

I would now like to turn from hadron production to muon production.[2] I do this because we find an anomalously large amount of muon production in noncoplanar two charged particle final states, and no anomaly in higher charged multiplicity states, when compared to QED and hadronic effects. The organization of my presentation will be to discuss first the two charged particle states, then the multi-charged particle states, and, finally, some interpretations of the results in terms of possible new particle production schemes.

The first events to be discussed are those having only two charged particles, one of which was tagged in the spectrometer as a muon with momentum greater than 1.05 GeV/c. These events were examined as a function of noncoplanarity as opposed to the non-collinearity which was used for the $e^+e^- \to \mu^+\mu^-$ events in Figure 2. The number of events versus the noncoplanarity angle, ϕ, is shown in Figure 8 for both $\sqrt{s}$ = 4.8 and 3.8 GeV. Usually, the second particle in these events penetrated the conjugate hadron filter and was identified as a muon. In some events, however, the second particle either missed the conjugate system, or failed to penetrate it. In either of these cases the identity of the particle was unknown. Events in which the second particle could not be identified are marked as "X" in Figure 8. In the $\sqrt{s}$ = 4.8 GeV data, for example, the 11 events with noncoplanarity less than $30°$ have a non-penetrating particle in the conjugate apparatus. Six of these have angles such that they do not pass through the back of the hadron filter. All are consistent with being minimum ionizing particles in the shower counter, which means that they could be muons, hadrons, or even electrons with energy below $\approx$ 400 MeV/c.

The curves in the figure represent the QED result from another program by Berends[6] et al. using the normalization as described earlier. Both distributions are in fair agreement with QED for $\phi < 20°$, but the $\sqrt{s}$ = 4.8 GeV data shows a clear excess of events above the QED curve at larger angles. The spectrometer muon momentum distribution of events for which $\phi > 20°$ at each s is also shown in Figure 8. Since hadrons are a potential source of muons it is noteworthy that, although lower momenta are favored (unlike $e^+e^- \to \mu^+\mu^-$), the distributions don't show as much momentum dependence as the hadron spectra in Figure 4.

The second class of events to be examined are defined by a muon with p > 1.05 GeV/c in the spectrometer and at least two additional particles in the "polymeter". No additional angular or momentum requirements are imposed upon these events because there is no large background source of muons in this channel. We found only two such events (n_{ch} = 3, 8) in the $\sqrt{s}$ = 4.8 GeV data.

The backgrounds from known sources of muons for each of these event types are listed in Table I for $\sqrt{s}$ = 4.8 GeV. The backgrounds related to hadron "punch through"[9] or decay were calculated using the momentum dependence of our measured hadronic spectra. The effects of the radiative tails of the ψ and ψ'

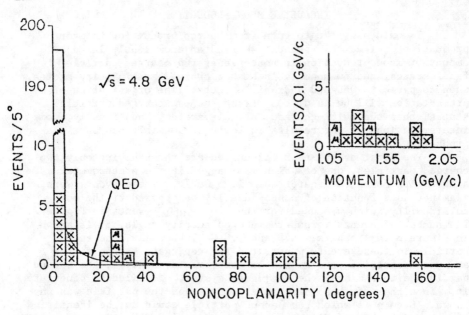
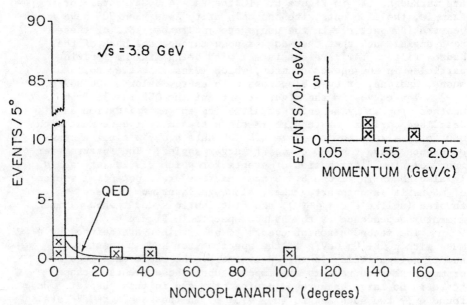

Fig. 8

MUON BACKGROUNDS AT $\sqrt{s} = 4.8$ GeV

	$n_{charge} = 2$ $\phi > 20°$	$n_{charge} > 2$
Hadron punch through	$.39 \pm .10$	$1.09 \pm .15$
$\pi \to \mu \nu$	$.52 \pm .13$	$1.77 \pm .22$
$K \to \mu \nu$	$.02 \pm .02$	$.14 \pm .09$
$e^+e^- \to \mu^+\mu^-\gamma$	2.95	–
$e^+e^- \to e^+e^-\mu^+\mu^-$	.01	2.05
$e^+e^- \to \psi'\gamma$ $\quad \hookrightarrow \psi \pi^+\pi^-$ $\qquad \hookrightarrow \mu^+\mu^-$	–	$.22 \pm .07$
$e^+e^- \to \psi'\gamma$ $\quad \hookrightarrow \psi(\text{neutral})$ $\qquad \hookrightarrow \mu^+\mu^-$	< .17	–
$e^+e^- \to \psi\gamma$ $\quad \hookrightarrow \mu^+\mu^-$	–	–
$e^+e^- \to \psi'\gamma$ $\quad \hookrightarrow \mu^+\mu^-$	–	–
Total number of events Expected	3.9	5.3
Observed	13	2

particles were found following the prescription of Jackson[10]. Muon production from the $\gamma\gamma$ process, $e^+e^- \to \mu^+\mu^- e^+e^-$, has been studied by Grammer and Kinoshita[11]. The specific calculations for this reaction in our apparatus, done by Grammer and Lepage, showed that mostly n_{ch} = 3 events would result. This is because of the noncoplanarity restriction, where the net transverse momentum for the muons must be balanced by the electrons. This causes one of the electrons to be deflected into the central detector, but not so much as to be seen in the conjugate shower counter.

The table shows that the total background in the n_{ch} = 2 and n_{ch} > 2 categories is 3.9 and 5.3 events respectively which is to be compared with observed 13 and 2 events. The observed n_{ch} > 2 rate is consistent with being entirely due to background, while there is a large excess of events in the n_{ch} = 2 data. The probability of the known processes explaining the n_{ch} = 2 events is about 2×10^{-4}. The cross section corresponding to the 9 extra events is

$$\left.\frac{d\sigma}{d\Omega}\right|_{90°, \phi>20°}^{n_{ch}=2} = 23^{+12}_{-9} \text{ pb/sr.}$$
$$p_\mu > 1.05 \text{ GeV/c}$$

If the extra muons are assumed to be produced isotropically, one finds

$$\sigma^{n_{ch}=2} = 285^{+150}_{-110} \text{ pb}$$
$$p_\mu > 1.05 \text{ GeV}$$
$$\phi > 20°$$

The upper limit on the cross section for n_{ch} > 2 muon production is

$$\left.\frac{d\sigma}{d\Omega}\right|_{90°}^{n_{ch}>2} < 7.5 \text{ pb/sr.}$$
$$p_\mu > 1.05 \text{ GeV/c}$$

at the 95% confidence level.

The n_{ch} = 2 data at $\sqrt{s}$ = 3.8 GeV gives

$$\left.\frac{d\sigma}{d\Omega}\right|_{90°}^{n_{ch}=2} = 17^{+32}_{-16} \text{ pb/sr.}$$
$$p_\mu > 1.05 \text{ GeV/c}$$

which is not a statistically significant signal.

No events with a μ-e signature were observed in this experiment. The conjugate shower counter would have identified such an event if the noncollinearity $< 40°$ and if $E_e > 400$ MeV/c. A necessarily crude and model dependent estimate using the SLAC-LBL results on μe production[12] suggests that we would have expected to see less than one event. Obviously, our seeing none is consistent with this estimate.

As a final comment on the numbers themselves, I'll point out that the background calculations for the $n_{ch} = 2$ and $n_{ch} > 2$ cases are clearly related. If one tries to increase the background in the $n_{ch} = 2$ case, the background in the $n_{ch} > 2$ case will probably also increase, quickly becoming inconsistent with the observed data.

INTERPRETATION OF MUON DATA

The SLAC-LBL[12] group has previously reported evidence for production of 2 charged particle states made up of $\mu^{\pm}e^{\mp}$ and have interpreted this as evidence for new particle production:

$$e^+e^- \to U^+U^-$$
$$\hookrightarrow e^-(\mu^-)X$$
$$\hookrightarrow \mu^+(e^+)X$$

If we follow this line of argument[13], we can place new limits on the U decay modes using our muon data without making any assumption about the nature (charm, heavy lepton, quark ...) of the U. If we assume that the spectrometer selects this type of event through the detection of the $\mu^{\pm}$ from the decay of the $U^{\pm}$ then the "polymeter" measures the total charged multiplicity of the $U^{\mp}$ decay. The $\sqrt{s} = 4.8$ GeV data can then be interpreted as giving

$$\frac{[U^{\pm} \to (n_{ch} \geq 3)]}{[U^{\pm} \to (n_{ch} = 1)]} = \frac{(< 7.5 \text{ pb/sr})}{(>23 \text{ pb/sr})} < \frac{1}{3}$$

with greater than 95% confidence.

We can compare this ratio to the predictions of models for different particle types. Three interesting possibilities are charm as described by Einhorn and Quigg[14], heavy leptons as described by Tsai[15], and quarks/gluons as described by Pati and Salam[16]. These models give values for this decay multiplicity ratio of > 2, $\approx .2$, and $\approx .2 \to .4$, respectively. The heavy lepton and quark/gluon model are consistent with the experimental data, whereas the charm model is quite inconsistent ($< .1\%$ C.L.) with the data. It can, therefore, be said that, whereas there may be charmed particle production in e^+e^- collisions at SPEAR, charmed particles are not the source of the anomalous muon events if the decay properties of charmed particles are as described by Einhorn and Quigg. Another way of saying this is that some new process other than (perhaps along with) charm production is occurring in e^+e^- reactions at $\sqrt{s} = 4.8$ GeV.

Looking specifically at the possibility of heavy leptons being the source of these anomalous muons, one can ask what branching ratio our data would suggest for $U \to \mu\nu\bar{\nu}$. This can be found using

$$\frac{d\sigma}{d\Omega}\bigg|_{e^+e^- \to U\bar{U}} = 23^{+12}_{-9} \frac{pb}{sr} = 2 \frac{d\sigma}{d\Omega} f_{1.05} g_{\phi>20°} B_1 B_{U \to \mu\nu\bar{\nu}}$$

where the 2 is due to our having an inclusive trigger, $\frac{d\sigma}{d\Omega}\bigg|_{e^+e^- \to U\bar{U}}$ is the average differential cross section for $e^+e^- \to U\bar{U} \to \mu$ near $90°$, $f_{1.05}$ is the average fraction of muons with momentum > 1.05 GeV/c, $g_{\phi>20°}$ is the correction for the noncoplanarity cut at $20°$, B_1 is the branching ratio of the second heavy lepton into one charged particle, and $B_{U \to \mu\nu\bar{\nu}}$ is the branching ratio of $U \to \mu\nu\bar{\nu}$. If we take the mass of the U to be 1.8 GeV, we can use the procedures developed in several papers[15,17] for calculating the heavy lepton production and decay factors $\frac{d\sigma}{d\Omega}\bigg|_{ee \to UU}$, $f_{1.05}$, and $g_{\phi>20°}$. We assume a V-A interaction and that the branching ratio, B_1, of U into one charged particle is $\approx .85$ as indicated by Tsai[15]. The result we get is

$$B_{U \to \mu\nu\bar{\nu}} = .23^{+.12}_{-.09}$$

which agrees with the SLAC-LBL result and theoretical predictions.

Our data can also be used to put a limit on muon production from charmed particles given several assumptions. The value of R, the ratio of hadron production to $\mu^+\mu^-$ production by e^+e^-, changes from about 2.3 to 4.8 as the energy changes from 3.0 to 4.8 GeV. If $\Delta R = 1$ is attributed to heavy lepton production, there remains an unexplained $\Delta R = 1.5$, or ≈ 5.2 nb, which could be due to charm production. If I assume that $e^+e^- \to c\bar{c}$ where c is any of the various possible charmed particles, and define B_μ as the branching ratio of $C \to K\mu\nu$ and $f_{1.05}$ is the fraction of the μ's with $P_\mu > 1.05$ GeV/c, then, very crudely, the inclusive cross section for muon detection is

$$\sigma(\mu^\pm \text{ detected}) = (5.2) \, 2 \, B_\mu \, f_{1.05} \text{ nb}$$

Since the charmed events will usually have $n_{ch} > 2$, we look at that cross section which, again, crudely is

$$\sigma_{(\mu^\pm)}^{n_{ch}>2} = \sigma_{(\mu^\pm)} [(1 - 2B_\mu)(B_3)]$$

where the $(1 - 2B_\mu)$ is the branching ratio into hadrons assuming

$B_\mu = B_e$, and B_3 is the probability of a hadron state having three or more charged particles. From Einhorn and Quigg[14] we roughly get $B_3 \approx 0.7$. If we assume that $M_c \approx 1.8 \to 2.0$ GeV than $f_{1.05} \approx .15$. We can then use our upper limit of

$$n_{ch} > 2$$
$$\frac{d\sigma}{d\Omega} < 7.5 \text{ Pb}$$

with isotropy in the above equation to find an upper limit for B of 12%. This is larger than theories predict, and can change with a different set of assumptions. It illustrates, however, that there is no inconsistency between charm production in e^+e^- and the absence of high multiplicity muon events in our experiment.

SP14/19 ENERGY SCAN NEAR 6 GeV

The last item in this talk is a description of an e^+e^- energy scan in the 5.7 to 6.1 GeV region searching for resonances. This was, of course, motivated by the announcement of the $T_{(5.97)}$[3]. The region was scanned in steps of $\approx$ 3 MeV with about 20 minutes being spent on each point. This generally gave an average $\int L dt$ of $10.7/_{nb}$. The equipment used in this run was different from that described earlier, so I'll describe it and the trigger, and then turn to the results.

A schematic of the SP14/19 central detector is shown in Figure 9. This figure shows three main components which need to be defined: 1) A set of "tube" proportional wire chambers surrounding the vacuum pipe, 2) four sets of three flat plane proportional wire chambers placed above, below, and on each side of the "tube" counters, and 3) two banks of NaI crystals, one above and one below the interaction region. The tube counters are each a one-wire PWC made by stretching one wire inside a long aluminum tube having a square cross section 1/2" on a side. The individual tubes were then packed side by side into four concentric rings, thus making a system equivalent to four concentric cylindrical PWC's. Their solid angle coverage was 90 → 95% of 4π steradians. The flat PWC's were standard chambers having their wires running perpendicular to the beam. These chambers gave a Θ measurement on charged particles whereas the tube counters gave a ϕ measurement. The flat chambers had a solid angle coverage of $\approx$ 25% of 4π steradians. The banks of NaI were made of a stack of several long ($\approx$ 18") hexagonally shaped (3" on a flat side, 6" "diameter") NaI crystals with a phototube on one of the hexagonal faces. The long axes of these crystals were horizontal and perpendicular to the e^+ or e^- beam direction. There were six crystals in one bank and seven in the other. The output signal of the NaI was fed to ADC's, but could also be used as part of the fast trigger.

The event trigger we used required at least two crudely defined charged particles in the tube counters along with either a charged particle or γ-ray in one of the NaI banks. This 3-fold coincidence gave a trigger rate of a few Hertz which is nearly 10^2 higher than the true e^+e^- interaction rate. Most of the triggers were background junk from particles being lost from the beam and from cosmic rays.

The events were computer analyzed to select "$e^+e^- \to$ hadrons"

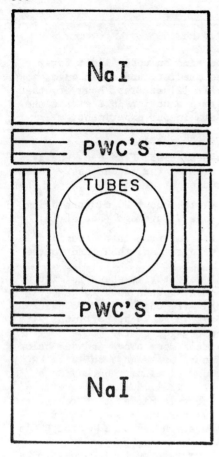

SP14/19
CENTRAL DETECTOR
Fig. 9

candidates from the junk. We generally required ≥ 2 charged
particles in the tube counters. An event having only two clear
tracks was classified as hadronic if the noncoplanarity angle
was greater than 120°. Events originating outside of the interaction region were rejected by imposing a 20 cm window on the reconstructed track origin along the beam direction for at least
one track.

The efficiency for detecting the decay products from a resonance depends markedly upon the final state characteristics. If
we assume that two charged particles are seen by the tube counters
and a neutral is converted in the NaI, the efficiency is

$$\text{Eff.} = (\Delta\Omega_T)^2 (\Delta\Omega_{NaI}) \, f_{(N,\theta,\phi)}$$

where $f_{(N,\theta,\phi)}$ is a function dependent upon the particle multiplicity and angular distribution for the event. (If one of the
charged particle in the tube counter triggers the NaI, one of the
$\Delta\Omega_T$ factors is removed, but then the $f_{(N,\theta,\phi)}$ doesn't include
neutral particles.) If we assume that the decay properties
of a resonance in the 5.7 → 6.1 GeV region is similar to the ψ or
ψ', this trigger efficiency is ≈ 60%.

Only preliminary results are available at this time, and
further work is being done. However, no significant bumps have
been observed in the data yet, which allows us to find a preliminary upper limit for the coupling constant of any missed resonance to e^+e^-. If we assume that the resonance is narrow, $\Gamma <$
3 MeV, it would show up in a couple of bins in the scan. Since we
don't see a significant fluctuation in a small number of bins, we
put a limit on the cross section integrated over this width of

$$\int \sigma_{(E)} dE = 40 \text{ nb-MeV}$$

This can be related to the coupling of a resonance to e^+e^- by

$$\int \sigma_{(E)} dE = \frac{6\pi^2}{m^2} \, \Gamma_{ee} \, \frac{\Gamma_h}{\Gamma_{total}}$$

where Γ_h is the width for decaying into hadron states detected by
the apparatus. If we assume the decay modes are similar to those
of the ψ or ψ, so that our efficiency is 60% on all hadron events,
then our preliminary upper limit is

$$\Gamma_{ee} < 100 \text{ ev}$$

for any undetected resonance, such as $T_{(5.97)}$, in this region.

SUMMARY

A summary of the main points I've tried to make in this talk
are as follows:

1) An anomalous muon production process exists in e^+e^-
reactions at 4.8 GeV.

2) It appears in our data only in two charged particle final

states.

3) If we assume that $e^+e^- \to U\bar{U}$, then

$$\frac{\Gamma(U \to (n_{ch} \leq 3))}{\Gamma(U \to (n_{ch} = 1))} < \frac{1}{3}$$

4) The above ratio is inconsistent with charm particles being the source of these muons, but charm particles could still be present in e^+e^- reactions in this energy region.

5) The decay multiplicity ratio is consistent with heavy leptons as is our total production rate.

6) Any narrow e^+e^- resonance in the $5.97 \to 6.1$ GeV region has $\Gamma_{ee} < 100$ ev.

REFERENCES

1. T. L. Atwood et al. Phys. Rev. Lett. $\underline{35}$, 704 (1975).
2. M. Cavalli-Sforza et al. SLAC-Pub. 1685, Dec. 1975, and Phys. Rev. Lett. $\underline{36}$, 558 (1976).
3. J. Appel's talk at this Conference. D. C. Hom et al., submitted to Phys. Rev. Lett., January 1975.
4. P. M. Joseph, Nucl. Instru. and Methods $\underline{75}$, 13 (1969); A. Buhler et al., Nuovo Cimento $\underline{35}$, 759 (1965).
5. J. E. Augustin et al., Phys. Rev. Lett. $\underline{34}$, 233 (1975); H. Newman et al., Phys. Rev. Lett. $\underline{32}$, 483 (1974); R. Madaras et al., Phys. Rev. Lett. $\underline{30}$, 507 (1973); B. L. Baron et al. Phys. Rev. Lett. $\underline{33}$, 663 (1974); B. Borgia et al., Lett. Nuovo Cimento $\underline{3}$, 115 (1972).
6. F. A. Berends, K. J. K. Gaemers, and R. Gastmans, Nucl. Phys. $\underline{B57}$, 381 (1973).
7. C. Morehouse, Proceedings of the SLAC Summer Institute on Particle Physics, July 1975, SLAC Report #191.
8. Private communication from K. J. K. Gaemers.
9. K. Abe et al. Phys. Rev. $\underline{D10}$, 3556 (1974). C. Jordan and Heinz-George Sander, Diplomarbeit, Aachen Univ. (W. Germany) (1974).
10. J. D. Jackson, LRL Memo No. JDJ/74-3 (unpublished).
11. G. Grammer, Jr. and T. Kinoshita, Nucl. Phys. $\underline{B80}$, 461 (1974).
12. M. L. Perl et al., Phys. Rev. Lett. $\underline{35}$, 1489 (1975).
13. I'm indebted to G. Snow for discussions on the interpretation of this data. Specifically, see G. Snow, Phys. Rev. Lett. to be published, Spring 1976.
14. M. B. Einhorn, C. Quigg, Phys. Rev. $\underline{D12}$, 2015 (1975).
15. Y. S. Tsai, Phys. Rev. $\underline{D4}$, 2821 (1971).
16. J. C. Pati, Maryland Report 76-071.
17. S. Pi, A. I. Sanda, Phys. Rev. Lett. $\underline{36}$, 1, 1976.

OTHER STRUCTURE IN e^+e^- ANNIHILATION*

Martin Breidenbach
Stanford Linear Accelerator Center, Stanford, CA 94305

ABSTRACT

The total cross section for the production of hadrons in e^+e^- annihilation exhibits complex structure in the region of center-of-mass energy about 4 GeV, where R, the ratio of the hadron to μ-pair cross sections, changes from a plateau value of about 2.5 to a new flat region of about 5.2. Corresponding structure is not seen in exclusive $2(\pi^+\pi^-)$ and $3(\pi^+\pi^-)$ cross sections, nor is structure seen in identified K^- spectra. Anomalous e-μ events are observed above an $E_{c.m.}$ of about 4 GeV, which are not explainable as arising solely from K^0 semileptonic decays of charmed mesons, or from two body decays of a new meson.

INTRODUCTION

The total cross section for the annihilation of e^+e^- into hadrons has yielded a rich structure even excluding the peaks of the $\psi(3095)$ and the $\psi'(3684)$. In particular there appears to be complex structure around $E_{c.m.}$ = 4 GeV, the region separating the domain where R, the ratio of hadronic to μ-pair cross sections, changes from a plateau value of around 2.5 to a new plateau around 5.2.

I will describe the newer results of the SLAC-LBL Magnetic Detector Collaboration[1] in the areas of total cross section measurements and in several areas where we have searched for other structure that might be associated with the transition from the "old" R $\approx$ 2 physics to the "new" R $\approx$ 5 physics. I will go over the status of the charm search, our K^- and $\bar{p}$ spectra, the information on exclusive multipion final states, and finally the status of the anomalous e-μ events. The apparatus has been described before, so I will only mention a few details as we go along.

We define a hadronic event as one with a vertex in the luminous region of the beams, having $\geq$ 3 prongs or two prongs acoplanar by $\geq 20°$, and having momenta > 300 MeV/c. The hadronic yields are corrected for backgrounds originating from beam gas or beam wall interactions, which are typically a few percent. The yields are normalized by measurements of Bhabha scattering in the detector, or, for the "fine scans", by measurements of small angle Bhabha scattering by small counters set into notches in the beam pipe. This is done, of course, to avoid the large statistical errors associated with the wide angle Bhabha scattering of these relatively short runs. The small angle luminosity measurements are calibrated and checked for consistency with the wide angle measurements.

The efficiency of the detector is determined from a set of Monte Carlo programs that estimate the probability of detecting p charged particles given that q were produced. Several models, varying from pure pion phase space to limited transverse momentum jets, were tried, with approximately 5% variations in the overall efficiency of the detector. The Monte Carlo generated probabilities were then used with observed charged particle distributions

---
*Work supported by the Energy Research and Development Administration.

to unfold the overall detector efficiency, which is shown in Fig. 1. The efficiency varies from about 30% at $E_{c.m.}$ = 2.5 GeV to about 65% at $E_{c.m.}$ = 8 GeV. The indicated errors in the total cross section plots, excluding the fine scans, are statistical with a 10% systematic error added in quadrature. We believe there may be an overall systematic error of 10% due to uncertainties in the normalization procedure, plus a possible $E_{c.m.}$ dependent error of about 15% varying smoothly with $E_{c.m.}$, due to errors in the determination of the detector efficiency.

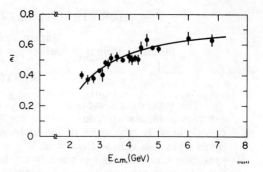

Fig. 1--Average detection efficiency vs $E_{c.m.}$.

The total cross section versus $E_{c.m.}$ of about a year ago is shown in Figure 2. The cross section due to the ψ and ψ' and their radiative tails has been removed, as has been done for the subsequent plots of σ_T and R. The cross section drops smoothly until encountering a broad peak around 4.1 GeV, and then resumes a smooth drop. Fig. 3 shows the total cross section at the time of the lepton-photon conference.[2] While the

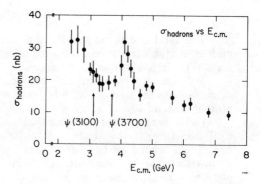

Fig. 2--Total hadronic cross section vs $E_{c.m.}$ circa January 1975.

general falling behavior of σ_T is unchanged, the area around 4.1 GeV seems more complex, indicating structure around 3.9 GeV and 4.4 GeV. A plot of R versus $E_{c.m.}$ corresponding to the data of Figure 3 is shown in Figure 4. R is flat at a value of about 2.5 up to $E_{c.m.} \approx 3.5$ GeV, roughly consistent with the value expected from colored u, d, and s quarks. R then goes through a complicated transition region and plateaus at a value of about 5.2. This contribution of the "new physics" seems high for only a charmed quark, and might indicate thresholds for heavy leptons or other new processes have been reached. (Note that most hadronic and semileptonic decay modes of a heavy lepton would satisfy the "hadron definition" of an event.)

Our most recent data for R between 3.8 and 4.6 GeV are shown in Figure 5. There is a broad peak between 3.9 and 4.3 GeV with indications of substructure. A small peak is seen at 3.95 GeV with a width of about 60 MeV, and a dip is seen near 4.08 GeV. Finally, another somewhat narrower peak is seen at 4.4 GeV. It is difficult to quantitatively obtain the parameters of these peaks. The resonances are occurring in the transition between the two plateau values of R. The threshold effects of the new channels that are opening up may distort Breit-Wigner line shapes. The shape of the

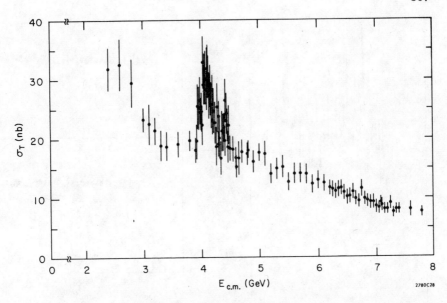

Fig. 3--Total hadronic cross section vs $E_{c.m.}$ at Lepton-Photon Symposium.

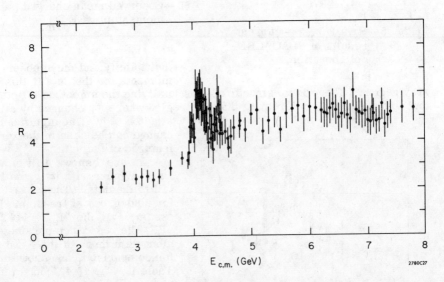

Fig. 4--R vs $E_{c.m.}$ corresponding to cross sections of Figure 3.

background is unknown so separation is problematical and the resonances may be interfering with the background and each other. We have fitted only the relatively separate peak at 4.4 GeV. Fig. 6 shows the mean charged

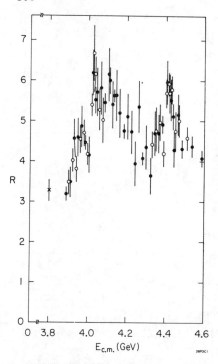

Fig. 5--R vs $E_{c.m.}$ - current results of SLAC/LBL collaboration.

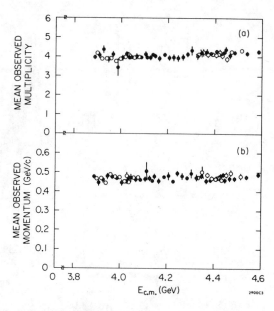

Fig. 6a--Observed mean charged particle multiplicity vs $E_{c.m.}$
6b--Observed mean charged particle momentum vs $E_{c.m.}$

energy region. There is no structure, thus justifying the smoothed efficiency that was used. It is interesting that in a region where σ_T changes by almost a factor of two, no significant change can be seen in these two features of the data.

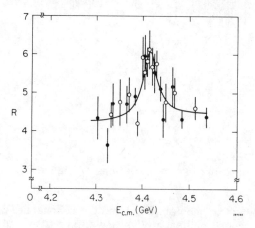

Fig. 7--R vs $E_{c.m.}$ for data used to fit $\psi(4414)$.

multiplicity and mean observed momenta for the data in this energy region.

Figure 7 shows a fit of a Breit-Wigner with its radiative tail to the data. The χ^2 is 17.9 for 20 degrees of freedom. If we assume J=1, then Γ_{ee} = 440 ± 140 eV. The resonance parameters, along with those of the ψ and ψ' for comparison, are shown in Table I. The $\psi(4414)$ has a full width more than a hundred times that of the ψ'; nevertheless its decay width into e pairs is around 1/5 that of the ψ'.

Shortly after the discovery of the ψ, SPEAR and the Magnetic Detector were set up to run

TABLE I

	$\psi(3095)$	$\psi'(3684)$	$\psi(4414)$
Mass (MeV/c^2)	3095 ± 4	3684 ± 5	4414 ± 7
Γ (MeV)	0.069 ± 0.015	0.228 ± 0.056	33 ± 10
Γ_{ee} (keV)	4.8 ± 0.6	2.1 ± 0.3	0.44 ± 0.14
B_{ee}	0.069 ± 0.009	0.0093 ± 0.0016	(1.3 ± 0.3) × 10^{-5}

in a scanning mode to search for other narrow resonances. The system's first (and only) success was the discovery of the ψ'. This scan and a subsequent scan up to $E_{c.m.}$ = 7.7 GeV are shown in Fig. 8. The first scan, excluding the ψ', sets upper 90% confidence limits on a narrow resonance of about 900 nb-MeV, while the high energy scan sets upper limits of about 450 nb-MeV. The errors shown are only statistical.

Due to the discovery in P-Be scattering of the Υ,[3] a peak at 5.97 GeV decaying into e$^+$e$^-$ pairs, there has been considerable interest in searching for it in e$^+$e$^-$ annihilation. Consequently, we conducted a search between 5.68 and 6.08 GeV with considerably greater sensitivity than the older fine scans. The preliminary online results are shown in Fig. 9. The integrated luminosity at each point corresponds to the production of about 20 μ pairs. The data are consistent with a constant value of R $\simeq$ 5.2. If the Υ is narrow compared to the SPEAR energy resolution of ~ 9 MeV FWHM (at $E_{c.m.}$ $\approx$ 6 GeV), then one can set an upper limit on Γ_{ee} of ~ 100 eV. If the Υ is wide enough to be resolved at SPEAR, then we set an upper limit on the branching ratio into e pairs, B_{ee}, of ~ 1 × 10^{-5}.

CHARM SEARCH

A search has been made in the invariant mass spectra of $K^\pm \pi^\mp$, $\pi^+\pi^-$, K^+K^-, $K_S K^\pm$, $K_S \pi^\pm$, $K_S \pi^+\pi^-$, $K^\pm \pi^\pm \pi^\mp$, and $\pi^+\pi^-\pi^\pm$, looking for peaks corresponding to a new meson. The amount of data that has been examined since our last publication[4] has approximately tripled, but the results are still negative. Data samples of roughly 10,000 events at $E_{c.m.}$ = 4.1, 4.4, and 4.8 GeV have been searched. If one assumes that the "new physics" is associated with the production of charmed mesons, then upper limits on various decay branching ratios can be set. These limits appear to "push" the expected values,[5] but do not rule out the theory. Of course, if part of the increase in R is due to phenomena other than charm, the upper limits are higher.

PARTICLE SPECTRA

The Magnetic Detector separates π's, K's, and p's by a time-of-flight system consisting of 48 plastic scintillators in a cylindrical array at a radius of 1.5 m from the beam. The time resolution is ~ 400 ps, allowing π-K separation up to momenta of 600 MeV/c and K-p separation up to 1.0 GeV/c. Only negative prongs are identified to avoid problems from beam-gas events which preferentially scatter protons. The particle spectra are

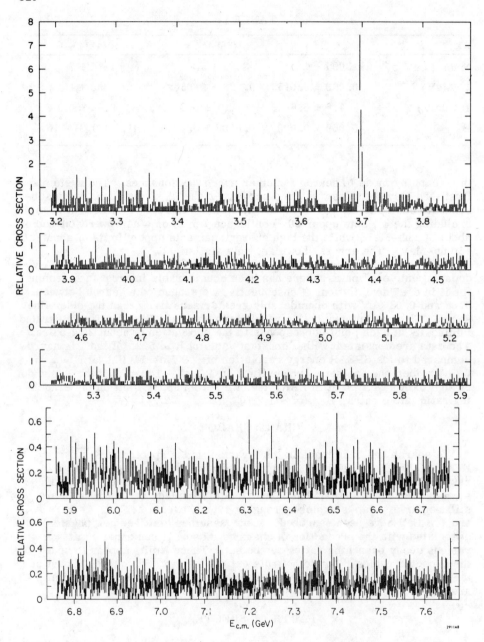

Fig. 8--σ_T vs $E_{c.m.}$ in fine steps over the range
a) 3.2 GeV $\leq E_{c.m.} \leq$ 5.9 GeV
b) 5.9 GeV $\leq E_{c.m.} \leq$ 7.6 GeV.

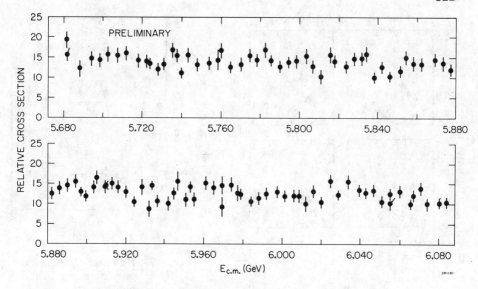

Fig. 9--High sensitivity fine scan of σ_T vs $E_{c.m.}$ for 5.68 GeV $\leq E_{c.m.} \leq$ 6.08 GeV.

corrected for trigger efficiencies, decay losses, and losses due to several particles striking a single timing counter.

Figure 10 shows the particle production cross sections at $E_{c.m.}$ = 4.8 GeV. The spectra at other energies are similar. The π's peak before the K's or p's. Figure 11 shows the fractions of K^- and $\bar{p}$ versus momenta for various $E_{c.m.}$ between 3.0 and 7.4 GeV. The fractions of K^- and $\bar{p}$ increase smoothly over the identified range in momentum. Incidentally, the K^- fraction must stop increasing by a momentum of 1 GeV/c in order to connect with the $K^{\mp}\bar{p}$ fraction of 0.21 ± 0.06 for particles with p > 1.1 GeV/c at $E_{c.m.}$ = 4.8 GeV measured by the Maryland-Pavia-Princeton group.[6] Fig. 12 shows the number of identified K^- and $\bar{p}$ per event versus $E_{c.m.}$. Since the identification procedure has fixed momentum cutoffs and

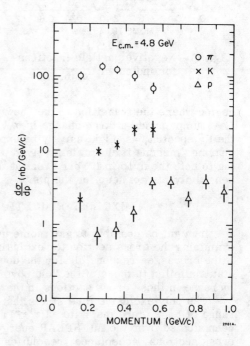

Fig. 10--$d\sigma/dp$ for π's, K's, and p's at $E_{c.m.}$ = 4.8 GeV.

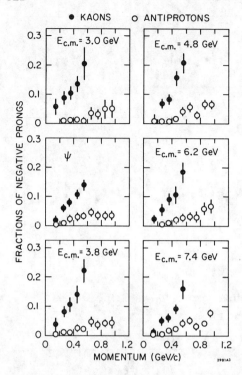

Fig. 11--Negative particle fractions for indicated $E_{c.m.}$

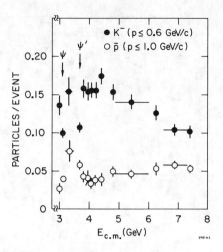

Fig. 12--Number of negative particles per event vs $E_{c.m.}$

the mean momentum of all charged particles is increasing with $E_{c.m.}$,[7] it is difficult to interpret the general behavior of the plot. However, we have previously seen that there is no dramatic change in the mean momentum of charged particles in the $E_{c.m.} \approx 4$ GeV region. It is in this region where the transition to the "new physics" occurs, and where the 4.1 GeV bump causes a large change in σ_T. Nevertheless, no change is seen in the K^- spectra, as might naively be expected from the decay products of charmed mesons if charm is involved with the step in R. It is also interesting to note the drop in K^- per event at the ψ and ψ'. This effect, with less sensitivity, is not seen for the $\bar{p}$'s.

EXCLUSIVE MULTIPION FINAL STATES

It might be possible to gain some understanding of the "new physics" by examining the cross section for exclusive final states as a function of $E_{c.m.}$. If the processes responsible for the doubling of R produced exclusive final states similar to those of the "old physics", we might expect a substantial increase in those cross sections in the $E_{c.m.} = 4$ GeV region.

The only exclusive final states that we are able to analyze over a wide range of $E_{c.m.}$ are the states $2(\pi^+\pi^-)$ and $3(\pi^+\pi^-)$. Other states have not been studied over the full SPEAR energy range because of limitations of cross sections, acceptance, resolution, and particle identification. The actual events analyzed can be those where all the prongs were seen (4-c events), or where one charged particle was missing (1-c events). 1-c events were not used above $E_{c.m.} = 4$ GeV to avoid contamination problems. The

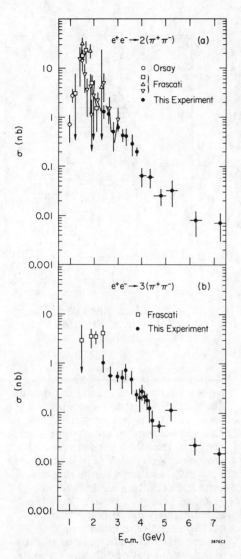

Fig. 13--Cross sections for exclusive production of a) $2(\pi^+\pi^-)$ and b) $3(\pi^+\pi^-)$ vs $E_{c.m.}$

data sample at 3 GeV was enriched by using data from ψ running. This is legitimate since the ψ has odd G-parity, implying that the decays to states with an even number of pions are mediated by a virtual photon.

The cross sections as a function of $E_{c.m.}$ are shown in Fig. 13. The upper plot is the $2(\pi^+\pi^-)$ cross section and is consistent with a smooth fall, $\sigma \sim s^{-2.8 \pm 0.5}$. The lower plot is the $3(\pi^+\pi^-)$ cross section and it also falls approximately like $\sigma \sim s^{-2.3 \pm 0.8}$. Neither set of data shows any structure in the $E_{c.m.} = 4$ GeV region.

At the low energy point, there are sufficient data to investigate resonance production within the exclusive states. Fig. 14 shows the invariant mass distribution for pairs of π's. In the $2(\pi^+\pi^-)$ data, strong ρ and f signals are seen; the solid curve is a fit using only uncorrelated Breit-Wigner shapes for the ρ and f, with a ratio of $\rho\pi\pi$ to $f\pi\pi$ of 1.9 ± 0.5. The $3(\pi^+\pi^-)$ data are fit with only a $\rho\pi\pi\pi\pi$ assumption. No f signal is seen. The dashed curves in both plots show invariant phase space. Figure 15 is a scatter plot of the mass of one pair of π's versus the mass of the other for the $2(\pi^+\pi^-)$ data, plotted so that $M_1 > M_2$. A clustering of points is seen at $M_1 \approx M_f$, $M_2 \approx M_\rho$, which is evidence for the exclusive process $e^+e^- \to \rho f$.

ANOMALOUS e-μ EVENTS

Probably by now most of you have heard the arguments for the anomalous e-μ signal seen at SPEAR. Rather than review these arguments,[8] I will discuss a small subset of the data from the so-called muon tower before going over the phenomenology of the e-μ data.

During the interval when SPEAR I changed into SPEAR II, two concrete absorbers corresponding to ~ 30 cm of iron were placed on top of the magnetic detector and spark chambers were installed above each absorber, as shown in Fig. 16. Candidate events for the tower analysis were required to have precisely two oppositely charged particles headed in directions so that,

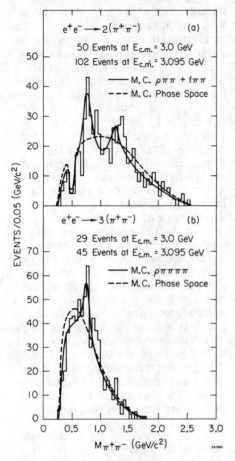

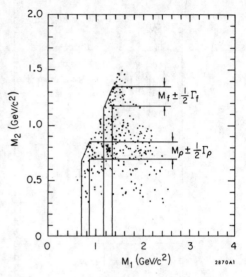

Fig. 14 -- $\pi^+\pi^-$ invariant mass distributions for a) $2(\pi^+\pi^-)$ and b) $3(\pi^+\pi^-)$ final states. The solid lines show fits using uncorrelated Breit-Wigner distributions for the indicated resonances plus uncorrelated pions. The dashed lines show invariant phase space.

Fig. 15 -- Invariant mass of $\pi^+\pi^-$ pairs for data of Fig. 14a ordered so that $M_1 > M_2$.

if they were muons, one should get to level 2 or level 3, and the other should get to a spark chamber at level 1. It was also required that no photons be seen in the shower counters; and that the charged tracks have momenta > 650 MeV/c, be acoplanar by $\geq 20°$, and have a missing mass squared recoiling against them greater than $(1.5 \text{ GeV}/c^2)^2$. The acoplanarity and missing mass requirements discriminate against radiative e^+e^- and $\mu^+\mu^-$ events. Fifty-eight events satisfying these requirements were found. A muon is identified as a particle with small pulse height in the shower counters and a signal within the expected area of the muon spark chambers. An electron is identified as a particle with large pulse height in the shower counter and no signal in the expected muon spark chamber area. Other prongs are called hadrons (h).

Misidentification probabilities for e's and μ's are determined from a sample of collinear lepton pairs. The probability that an electron simulates a μ at levels 1 and 2 in the tower is less than 2×10^{-3}. The probability that a μ gives a large shower counter signal and does not penetrate the absorber

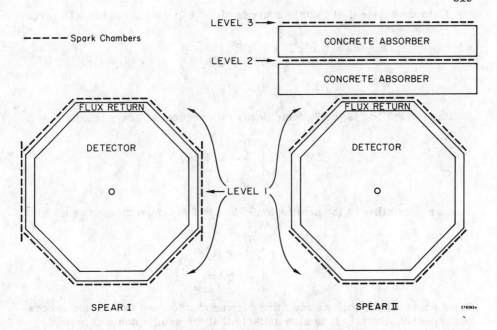

Fig. 16--Muon detectors in SPEAR I and SPEAR II.

is less than 3×10^{-3}. Hadron misidentification probabilities are estimated from data in which 3 or more prongs are detected and all are assumed to be hadrons. The probability of a hadron simulating a μ at level 2 is about 7% and the probability of a hadron simulating an e is about 20%.

Of the 58 events, 10 were identified as e^+e^-, 11 as $\mu^+\mu^-$, and 37 as combinations of e, μ, and h, including 5 $e\mu$ with the μ identified at level 2 or 3. A conservative estimate of the misidentification background can be made by assuming the 37 events are all hh. The arithmetic is summarized in Table II,

TABLE II

Event Type	Number	Misidentification Probability	Background Contribution
ee	10	0.002	0.02
$\mu\mu$	11	0.003	0.03
hh	37	0.2×0.07	0.53
TOTAL	58		0.57

and the expected background is 0.57 $e\mu$ events. The statistical probability that the 5 observed $e\mu$ events are due to the background is about 3×10^{-4}.

Onto the physics! Possible sources of the e-μ events might be the decay of a new meson:

$$e^+e^- \to U^+U^-$$
$$\hookrightarrow e^-\bar{\nu}_e$$
$$\hookrightarrow \mu^+\nu_\mu$$

Another possibility is the leptonic decay of a sequential heavy lepton:

$$e^+e^- \to U^+U^-$$
$$\hookrightarrow e^-\bar{\nu}_e\nu_U$$
$$\hookrightarrow \mu^+\nu_\mu\bar{\nu}_U$$

Another possibility is the semileptonic decay of charmed mesons, such as:

$$e^+e^- \to U^+U^-$$
$$\hookrightarrow K^0 e^-\bar{\nu}_e$$
$$\hookrightarrow \bar{K}^0\mu^+\nu_\mu$$

Other possibilities such as radiative pair production and two-photon processes appear unlikely.[8] It is also unlikely that the semileptonic decays of charmed mesons are the sole source of the e-μ events. The $\pi^+\pi^-$ signal of a K_s decay is clearly observed in the magnetic detector. In a data set having 49 eμ events, a search was made for events of the form $e\mu K_s$, eeK_s, and $\mu\mu K_s$. If all of the e-μ events were from K^0 semileptonic decays, then, putting in the efficiencies and acceptances, 47 $\ell\ell K_s$ events should have been seen. None were observed. Another way to state the result is that the fraction of eμ events due to K_0 semileptonic decays is less than 5% at a 90% confidence level. If other semileptonic decay modes are considered, the upper limit to the semileptonic contribution is less than 19 to 32%, depending upon assumptions about the misidentification of e-μ events.

The raw cross section for the production of eμ events versus $E_{c.m.}$ is shown in Fig. 17. The cross sections have not been corrected for the detector acceptance or the kinematic cuts since the origin of the events is unknown. There appears to be a threshold slightly below $E_{c.m.} = 4$ GeV. The statistical accuracy is insufficient to distinguish between the production of a heavy lepton pair, which would fall like β/s, or meson pairs, which would fall like $(\beta/s)^3$. More information can be gained by combining the data at different $E_{c.m.}$ to examine the

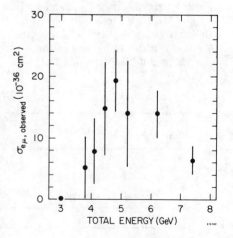

Fig. 17--Uncorrected cross section for observation of e and μ and no other particles vs $E_{c.m.}$.

317

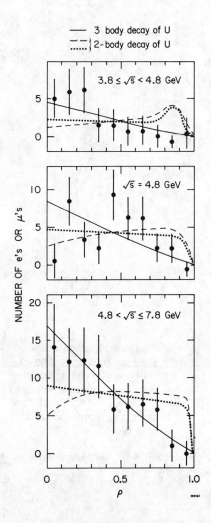

Fig. 18--Distribution in $\rho = (1-0.65)/(p_{max}-0.65)$ for all $E_{c.m.}$. The solid curve is the distribution expected for the decay of 1.8 GeV/c² heavy leptons. The dotted curve represents isotropic decays of a 1.9 GeV/c² boson. The dashed curve is similar to the dotted curve except the collinearity angle distribution which has been set to fit the data.

momentum spectrum by constructing a variable

$$\rho = \frac{p - 0.65 \text{ GeV/c}}{p_{max} - 0.65 \text{ GeV/c}}$$

The data are shown in Fig. 18. The solid line is the distribution expected from the V-A decays of heavy leptons of mass 1.8 GeV/c². It is a good fit with a χ^2/DOF of about one. The dotted curve is the distribution expected from an isotropic two-body decay while the dashed curve is a two-body decay with a collinearity angle distribution adjusted to match the data. Both curves are poor fits to the data; the isotropic decay curve has a χ^2/DOF of about four and the curve which matches the collinearity distribution is worse. This indicates that not all of the e-μ events come from two-body decays. Figure 19 shows the collinearity angle distributions of the e-μ's for three ranges of $E_{c.m.}$. The heavy lepton distributions are again better fits than the two-body decay distributions. The distribution becomes much less isotropic as $E_{c.m.}$ increases, which is characteristic of the production of a pair of particles.

SUMMARY

The total hadronic cross section to μ pair ratio R is flat above and below a transition region around 4 GeV. The transition region has at least three peaks in it. No other structure is seen in the total cross section, nor is any seen in K⁻ spectra, or in 4π and 6π exclusive cross sections. There are no encouraging results from the charm search, but the present limits are not very damaging to charm theories. Anomalous e-μ events exist, with origins that cannot be explained as coming exclusively from semileptonic decays or two-body decays.

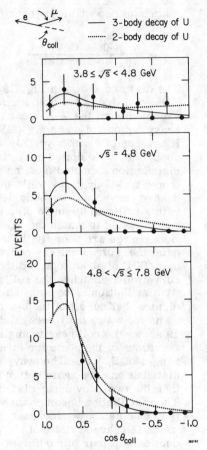

Fig. 19--The distribution of $e\mu$ events vs the cosine of their collinearity angle for three different ranges of $E_{c.m.}$. The curves have the same meaning as in Fig. 18.

Hopefully some of the questions implied by these data will be answered by the next conference.

REFERENCES

1. The members of the SLAC/LBL collaboration: G. S. Abrams, J. E. Augustin, A. M. Boyarski, M. Breidenbach, D. Briggs, F. Bulos, W. Chinowsky, G. J. Feldman, G. E. Fischer, C. E. Friedberg, D. Fryberger, G. Goldhaber, G. Hanson, D. L. Hartill, R. J. Hollebeek, J. Jaros, B. Jean-Marie, J. A. Kadyk, R. R. Larsen, A. M. Litke, D. Lüke, B. A. Lulu, V. Lüth, H. L. Lynch, R. Madaras, C. C. Morehouse, K. Nguyen, J. M. Paterson, M. L. Perl, I. Peruzzi, F. M. Pierre, M. Piccolo, T. P. Pun, P. Rapidis, B. Richter, B. Sadoulet, R. F. Schwitters, J. Siegrist, W. Tanenbaum, G. H. Trilling, F. Vannucci, J. S. Whitaker, F. C. Winkelmann, J. E. Wiss, and J. E. Zipse.
2. R. F. Schwitters in Proceedings of the International Symposium on Lepton and Photon Interactions at High Energies, Stanford, California,

1975, edited by W. T. Kirk (Stanford Linear Accelerator Center, Stanford, California, 1975).
3. D. C. Hom et al., "Observation of High Mass Dilepton Pairs" (to be published).
4. A. M. Boyarski et al., Phys. Rev. Lett. $\underline{35}$, 196 (1975).
5. M. K. Gaillard, B. W. Lee, and J. L. Rosner, Rev. Mod. Phys. $\underline{47}$, 277 (1975).
6. T. L. Atwood et al., Phys. Rev. Lett. $\underline{35}$, 704 (1975).
7. J. E. Augustin et al., Phys. Rev. Lett. $\underline{34}$, 764 (1975) and Ref. 2.
8. M. L. Perl et al., Phys. Rev. Lett. $\underline{35}$, 1489 (1975) and M. L. Perl, SLAC Report No. SLAC-PUB-1592 (1975) to be published in Proc. Canadian Institute of Particle Physics Int. Summer School, McGill University, Montreal, Canada, June 16-21, 1975.

POSTLUDE*

C. Quigg[†]

Fermi National Accelerator Laboratory,[§] Batavia, IL 60510

ABSTRACT

A number of comments are made on issues raised by experimental results presented to the Conference. Prejudices and disconcertions are shared openly.

INTRODUCTION

Rather than attempt a summary of what we have heard or of where our field stands, I want to share with you my reactions to the findings reported here. These will include expressions of delight and of bewilderment, clarifications of folklore, and theoretical background comments. I have made little effort to achieve uniform coverage or to present all sides of every question. Indeed, what follows is simply my critical reading — here provocative, there didactic, elsewhere intemperate — of the papers delivered here.

HADRONIC PRODUCTION OF NEW PARTICLES

A number of searches for charmed or otherwise novel particles produced in hadron-hadron collisions have been carried out.[1-7] These have yielded a few curiosities but no one yet has a compelling candidate for the wallet cards. What is the interesting level of sensitivity for charm searches in hadron-hadron collisions? My viscera, and those of my colleagues[8,9] say that at Fermilab energies the total cross section for charm production is

$$\sigma(\text{charm}) \sim 1 \ \mu b \ .$$

If we divide this number by 3 for the distinct species of charmed pseudoscalars and multiply by 5% which appears to be a reasonable guess[10] for a typical nonleptonic branching fraction, we arrive at

$$\sigma \times \text{Branching ratio} \sim 1 \text{ to } 10 \text{ nb}$$

---

*Concluding remarks at the II International Conference on New Results in High Energy Physics, Vanderbilt University, March 1-3, 1976.

[†]Alfred P. Sloan Foundation Fellow; also at Enrico Fermi Institute, University of Chicago, Chicago, Illinois 60637

[§]Operated by Universities Research Association Inc. under contract with the Energy Research and Development Administration.

as the interesting level to penetrate. This is so small
that the unsuccessful searches do not yet surprise or dismay
the advocates of charm.

Special triggers have been employed to select events
in which charmed particle production may be enhanced. If
charmed particles have appreciable semileptonic decays, a
(prompt) μ trigger might select events in which one charm
has decayed semileptonically, leaving the other to be observed.
The sensitivity required for this kind of search
is no less demanding than in the untriggered case. With
a semileptonic branching fraction of approximately 10%, we
may expect to require

$$\sigma \times \text{(Semileptonic branching ratio)}$$
$$\times \text{(Branching ratio for decay of second charm)}$$
$$\lesssim 1 \text{ nb} .$$

The use of a ψ trigger[11] to enhance the charm signal to
background has also been attempted. Here it is assumed
that ψ is a $(c\bar{c})$ bound state produced in an Okubo-Zweig-
Iizuka rule[12] — respecting exchange process such as

$$\pi \quad \bar{D} \quad \psi \quad D \quad N$$
$$\pi \quad N$$

where solid lines represent ordinary light quarks and
broken lines represent charmed quarks. According to this
line of reasoning, every ψ is accompanied by a pair of
charmed particles. Likewise, the replacement of strange
quarks for charmed quarks in the exchange diagram leads
to the expectation that every produced ϕ should be accompanied
by a pair of strange particles. A search[13] in
low energy pp collisions has produced no evidence for
a $\phi(K\bar{K})$ enhancement. What is wrong[14] with the argument?
First, the OZI rule is unlikely to be exact; ϕ and ψ
do decay into nonstrange and uncharmed hadrons, respectively.
Second, it is not the only influence on production
cross sections. For example, in 24 GeV/c pp collisions
a final state like $pK\phi\Lambda$ is energetically far more costly
than $p\phi p$. Finally, it may be that the dominant mechanism
for ψ production is OZI-rule violating and has nothing
to do with charmed particle exchange. In this event there
is no reason to expect any $\psi(D\bar{D})$ correlation whatsoever.

We meet such a mechanism by asking whether ψ' is produced in hadron-hadron collisions. Data presented to this conference[15,16] indicate that in 300-400 GeV/c NN collisions

$$\sigma(\psi')/\sigma(\psi) \lesssim 1/10 \quad .$$

Earlier, the MIT-BNL group reported[17]

$$\sigma(\psi')/\sigma(\psi) < 1/100$$

in 28.5 GeV/c pBe collisions. Regardless of the numbers, no one has yet seen a clean ψ' peak in hadronic interactions.[18] Is this amazing? How might it be explained? Three possibilities leap instantly to mind. The first is a kinematical suppression.[19] I find this unappealing at several hundred GeV/c, but we may entertain a thermodynamic argument anyway. According to the usual lore[20] the ratio of production cross sections for particles whose masses differ by ΔM is, <u>ceteris paribus</u>, $\exp[-\Delta M/160 \text{ MeV}]$. For the case at hand, we have

$$\sigma[\psi'(3684)]/\sigma[\psi(3095)] = 0.025 \quad ,$$

which leaves us with nothing to explain. My only objection to arguments of this kind is that I do not understand them. A second coneivable explanation is that ψ and ψ' are not closely related objects. I dismiss this at once.[21] The third possibility is the one I find most interesting, if not entirely convincing: it is that the suppression has a specific dynamical origin[22] suggested by the charmonium picture of Appelquist and Politzer.[23] As Appelquist has reminded us in his talk,[24] the charmonium-positronium analogy explains the narrowness of ψ and ψ' by the requirement that these C=-1 $^3S_1$ states decay through 3-gluon intermediate states, whereas C=+1 psions (such as $^1S_0$ η_c and the χ or P_c states) can decay through 2-gluon intermediate states and should be broader. On this argument, the total width of the pseudoscalar η_c is expected to be about 5 MeV, 75 times the width of ψ. Let us now apply this reasoning to the OZI rule violating production mechanism posited by Einhorn and Ellis,[9] a Drell-Yan[25] process for gluons. Evidently

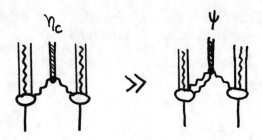

(the wavy lines represent gluons), so that

$$\sigma(\chi) \text{ or } \sigma(\eta_c) \gg \sigma(\psi) \text{ or } \sigma(\psi') \quad .$$

Taking this line of reasoning too literally we arrive at the conclusion that <u>no ψs or ψ's are produced directly in hadron-hadron collisions</u>. All the ψs we see occur as decay products in reactions like

$$\pi N \to \chi + \text{anything}$$
$$\hookrightarrow \psi + \gamma + \cdots$$

Notice that because $\chi(3500) \not\to \psi'(3684) + \cdots$, we require more massive C=+1 psions to feed ψ' production. If these lie below the putative charm threshold, E1 transitions to ψ should be favored over those to ψ' by the larger Q-value. If they lie above the threshold, they may not decay appreciably into either ψ or ψ'. In either instance we have reason to expect

$$\sigma(\psi) \gg \sigma(\psi') \quad .$$

It is important to put this proposal to the test by searching for the decay photons accompanying ψs and by searching directly for χ production in reactions such as

$$\pi N \to \chi + \text{anything}$$
$$\hookrightarrow \pi^+ \pi^- \quad .$$

Unfortunately, the branching ratios I glean from Friedberg's report[26] do not encourage the hope that this will be easy.

HADRONIC PRODUCTION OF LEPTONS

One may ask whether ψ is produced strongly, or through the intermediary of a virtual photon. An old but good argument against the latter possibility is that a direct (i.e., nonelectromagnetic) coupling of ψ to hadrons is implied by the observation that

$$\Gamma(\psi \to \text{hadrons}) \gg \Gamma(\psi \to \ell^+\bar{\ell}) \frac{\sigma(e^+e^- \to \text{hadrons})}{\sigma(e^+e^- \to \mu^+\mu^-)} \bigg|_{\text{off resonance}}.$$

A less devious proof can be had by verifying that ψ production respects strong interaction symmetries such as isospin invariance. The ratio

$$\frac{\sigma(\pi^+ C \to \psi + X)}{\sigma(\pi^- C \to \psi + X)} = 1$$

for strong production, including the cascade mechanism reviewed in the preceding section. I believe we know the outcome of this measurement before it is done: the ψ is produced strongly. However, for lepton pairs in general the origin and the production mechanism are quite uncertain. The same test applies.[27] We define

$$\rho = \frac{d\sigma}{dM^2}(\pi^+ C \to \ell^+\ell^- + X) \bigg/ \frac{d\sigma}{dM^2}(\pi^- C \to \ell^+\ell^- + X) ,$$

where M is the invariant mass of the $\ell^+\ell^-$ pair. If the pairs are produced through

$$\pi^\pm C \to \text{hadron} + X$$
$$\hookrightarrow \ell^+\ell^- ,$$

$\rho=1$, but if they are produced through

$$\pi^\pm C \to \gamma_V + X$$
$$\hookrightarrow \ell^+\ell^- ,$$

$\rho \neq 1$ in general. An extreme example of the latter is Drell-Yan production by valence quark annihilation. For incident π^+, the elementary process is $d\bar{d} \to \gamma_V$; for incident π^- it is $u\bar{u} \to \gamma_V$. The cross section is proportional to the square of the quark charge, so we find

$$\rho = \tfrac{1}{4} .$$

In a less schematic Drell-Yan model, with valence and sea quarks, the value $\rho=\frac{1}{4}$ is attained only for rather large lepton pair masses. At very small masses, for which sea quark-sea quark annihilations are dominant, $\rho \to 1$. This is shown in Fig. 1 for two choices[28,29] of parton distributions.

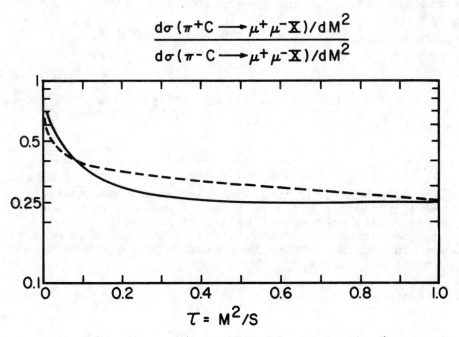

Fig. 1: Ratio of lepton-pair production in $\pi^{\pm}C$ collisions expected in the Drell-Yan model. The solid line is based on the parton distributions of Ref. 28; the dashed line on those of Ref. 29.

Is there indeed any lepton-pair continuum? Is it of Drell-Yan origin? Sanders,[30] reporting on a Chicago-Princeton experiment at Fermilab, has raised the possibility that there may be no continuum at all between $\frac{1}{2}$ and 3 GeV/c^2, that vector mesons $\rho, \omega, \phi, \rho'$ might account for the entire observed signal. Increased resolution and the isospin test just discussed will help to clarify the situation. In his report on the Columbia-Fermilab-Stony Brook experiment, Appel[16] indicated that perhaps there is a continuum contribution above 5 GeV/c^2. If you believe $T(5.97$ GeV/c$^2)$ is a resonance, the evidence for a continuum is very weak indeed. The extant data cannot sustain both a prominent resonance and a continuum. I cannot agree with Sullivan's conclusion[31] that the Drell-Yan mechanism is firmly established. We still require the convincing demonstration that $M^4 d\sigma/dM^2$ depends only upon M^2/s, that the pairs originate from virtual photons as distinct from vector mesons, and the observation of a credible continuum signal for values of M^2/s which are not tiny. Defensible estimates[32] of the Drell-Yan continuum lie at or below the levels reported by Appel[16] and by Sanders.[30] The identification of the Drell-Yan signal is of great importance for discussions of new accelerators intended as W boson factories.

At low lepton pair masses where we cannot expect Drell-Yan arguments to apply, the experimental situation is even more confusing — at least to me. There are backgrounds from η Dalitz pairs and from Bethe-Heitler conversion of the photons released in π^0 decay. Evidently both of these are imperfectly understood in practice. The rate for $\eta \to \mu^+\mu^-\gamma$ is only known theoretically, and different reasonable choices for the form factor lead to predictions[33] which differ by as much as 30%. Much more work, principally on low mass e^+e^- pairs, will be required to assess theoretical speculations[34] that copious radiation of nearly-real photons is responsible for the bulk of prompt lepton production.

The reports[15,16,35-39] on prompt production of leptons leave me terribly perplexed. I think this is not due entirely to my hebetude, but to contradictions among data sets. It appears to me that we speak with confidence about μ/e, μ/π, and e/π when in fact very little is known about the dependence of these ratios upon s, x, and $p_\perp$ beyond what was reported in 1974.[42] I simply don't know what to believe at small $p_\perp$ or at low energies. I should prefer to have the differences between experiments faced directly and not disregarded.

CHASING DOWN CHARM

Do charmed particles exist? Beyond reasonable doubt. Have they been discovered? Almost certainly. By whom? Ask me in a year.

I want to issue a warning about charmed particle mass formulas, which seem to be taken very seriously in some quarters. If you are told that someone can compute charmed particle masses to within a few MeV/c^2, my advice is to get a firm grip on your wallet. There are several ways of estimating these masses.[43] All are equally (in)credible. The situation is not closely analogous to the propitious days of the hunt for Ω^-, and experimental agreement with a theoretical mass is of far less consequence.

Although I am unprepared to state who has discovered charmed particles, we have been assured[44-46] that it isn't the e^+e^- annihilators. Why not? One possibility is that the quark of which psions are made is not the charmed quark. We lack any indication of a connection between the constituents of ψ and the weak interaction, but let us tentatively put aside this possibility in the interest of simplicity and because the charmed quark we believe we need[8] ought to lie in the explored mass region. More to the point is that we don't know how many new phenomena[47] are occurring in the range $3 \text{ GeV} < s^{1/2} < 5 \text{ GeV}$. As confidence grows[44,48] in the reality of the heavy leptons reported by Perl and coworkers at SPEAR,[49] the limits on charmed meson production recede from the threatening levels[50] of last summer.[51] Despite my faith in the reliability of these excuses, I shall feel more confortable when they no longer have to be made.

ISSUES IN NEUTRINO-INDUCED PRODUCTION OF CHARMS

In the Glashow-Iliopoulos-Maiani (GIM) scheme,[52] the hadronic weak current mediated by W^+ has the form

$$J^+ \sim \bar{u}(d \cos\theta_C + s \sin\theta_C)$$

$$+ \bar{c}(s \cos\theta_C - d \sin\theta_C) \quad .$$

The observation of $K_s e^+ (\nu) \mu^-$ events in the 15' chamber at Fermilab[53] and in the Gargamelle chamber at CERN[54] hints that the GIM current may indeed by operating. It was anticipated[8] that of the charmed pseudoscalar mesons $D^+ (= c\bar{d})$ would be most likely to have appreciable semileptonic decays. The simplest semileptonic decay mode is

$$D^+ \to \bar{K}^0 \ell^+ \nu \quad ,$$

which invites identification with the observed events. Two questions of vital importance are whether the $K_s e^+ \mu^-$ events truly signal charm, and how the apparent new objects are produced. At the moment we can answer neither of these incontrovertibly. I shall therefore indicate some of the processes whereby charmed particles may be produced in $(\nu, \bar{\nu})N$ collisions.[55]

Let us first consider charged current interactions. The most reasonable possibilities for charm production in νN collisions are indicated in Fig. 2. The Cabibbo-favored diffractive process[56] $\nu N \to \mu^- + F^{*+}$ + anything likely results in the semileptonic decay

$$F^+ \to (\ell^+ \nu)(s\bar{s})$$
$$\hookrightarrow \eta^0 \text{ or } K\bar{K} \quad .$$

It is a "delayed threshold" reaction in which the total hadronic energy must be large enough that the four-momentum transfer squared required to put the F^* on

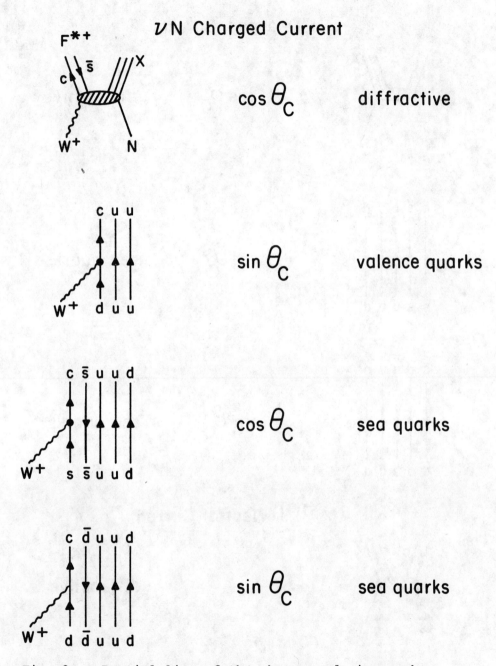

Fig. 2: Partial list of the charm-producing mechanisms which may operate in $(\nu,\bar{\nu})$-nucleon collisions. See Ref. 55 for additional discussion.

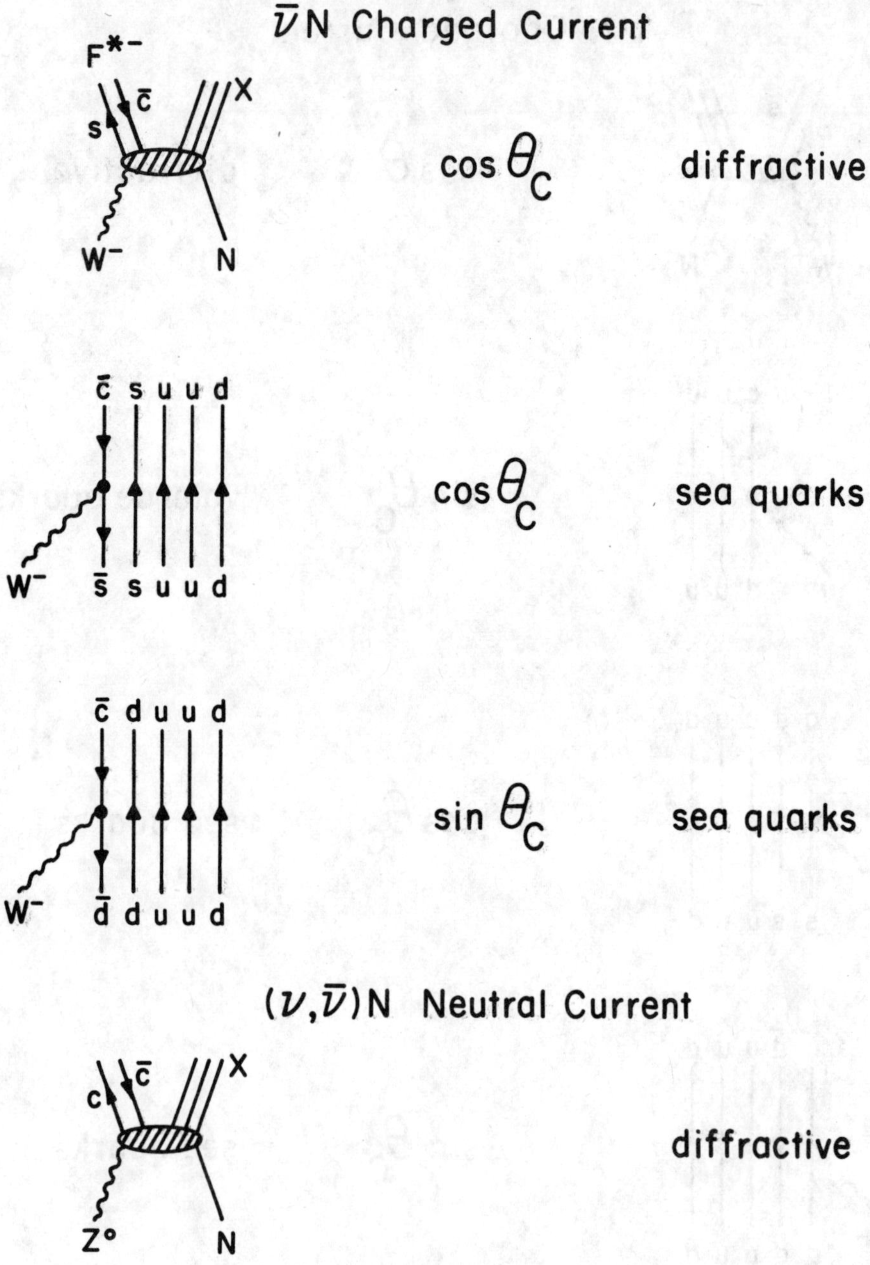

ν̄N Charged Current
F*−
s
c̄
X
W−
N
cos θC
diffractive
c̄ s u u d
s̄ s u u d
W−
cos θC
sea quarks
c̄ d u u d
d̄ d u u d
W−
sin θC
sea quarks
(ν,ν̄)N Neutral Current
c
c̄
X
Z°
N
diffractive

Fig. 2 (continued):

mass shell is small. The other reactions indicated lead to final states containing a single charmed meson or baryon, plus anything else. Some examples of expected semileptonic decays are[57]

$$D \to \bar{K}\ell^+\nu$$

$$C_1 \to \Sigma\ell^+\nu \quad .$$

Antineutrino-induced collisions are distinguished in two important ways from neutrino-induced collisions: No charm-changing interactions with valence quarks can take place, and no charmed baryons will be produced. The remaining mechanisms for the production of charmed particles, shown in Fig. 2, are counterparts of those we have already mentioned in the νN case.

Opportunities for charm production are more restricted in neutral current interactions. The neutral current can be represented schematically by

$$J^0 \sim u\bar{u} + c\bar{c} - d\bar{d} - s\bar{s} \quad ,$$

which gives rise to the diffractive production of charmed particle pairs by the last diagram in Fig. 2. Because two (heavy) charmed particles are produced by this process, I expect it to have an effective threshold energy about four times that required for diffractive F^* production. Following Einhorn and Lee[56] I estimate the needed beam energy at about 120 GeV.

PSION SPECTROSCOPY

The observation[26] of the transition

$$\psi' \to \gamma(260 \text{ MeV}) + X$$

at the 8-10% level removes a principal embarrassment for charmonium spectroscopists.[24,58] The remaining great problem is the mass of the apparent 0^{-+} state[44,45,59] $\eta_c(2800)$. With theoretical success comes theoretical hubris. Theoretical hubris wants the η_c mass to lie between 3000-3050 MeV/c^2. It would be delightsome if nature could be persuaded to cooperate.

We learned from Breidenbach[44] that the SLAC-LBL Group has failed to confirm the existence of $T(5.97? \text{ GeV/c}^2)$[60] in the mass interval between 5.8 and 6.1 GeV/c^2. It is conceivable the T is produced copiously in pp collisions but not in e^+e^- annihilations if it is not a vector particle, or if its branching fraction into lepton pairs is tiny. To my mind it is too early to be very quantitative, but one should be prepared to be put off by the implication of a large production cross section (i.e., much greater than that of ψ) in hadronic interactions. Another argument discourages the belief that Υ represents the ψ-analog of yet another quark. If it were a bound state of a new heavy quark, we should expect a step in

$$R \equiv \frac{\sigma(e^+e^- \to \text{hadrons})}{\sigma(e^+e^- \to \mu^+\mu^-)}$$

within about 1 GeV. The step need not be as spectacular as the one near 4 GeV, where several new phenomena may coincide. For a quark of charge 1/3, the increment in R would be only 1/3, but there is no evidence[44] for even such a small change above $\sqrt{s}$ = 5 GeV. I have the same uneasiness about a heavy quark interpretation of the noname (4.8 GeV/c^2).[15]

ACKNOWLEDGEMENTS

It is a pleasure to thank Bob Panvini and his Vanderbilt colleagues for their gracious southern hospitality. Special thanks go to Steve Ellis and Lem Motlow for their help in preparing these remarks.

FOOTNOTES AND REFERENCES

1. Min Chen, These Proceedings.
2. E. Shibata, These Proceedings.
3. N. McCubbin, These Proceedings.
4. R. Harris, These Proceedings.
5. D. Ritson, These Proceedings.
6. H. Lubatti, These Proceedings.
7. J. Matthews, These Proceedings.
8. M.K. Gaillard, B.W. Lee, and J.L. Rosner, Rev. Mod. Phys. 47, 277 (1975).
9. M.B. Einhorn and S.D. Ellis, Phys. Rev. D12, 2007 (1975); Phys. Rev. Lett. 34, 1190 (1975).
10. M.B. Einhorn and C. Quigg, Phys. Rev. D12, 2015 (1975).
11. D. Sivers, Phys. Rev. D11, 3253 (1975).
12. S. Okubo, Phys. Lett. 5, 165 (1963); G. Zweig, oral tradition (1964); J. Iizuka, Prog. Theor. Phys. Suppl. 37-38, 21 (1966).
13. V. Blobel, et al., Phys. Lett. 59B, 88 (1975).
14. H.J. Lipkin, Phys. Lett. 60B, 371 (1976).
15. I. Gaines, These Proceedings.
16. J. Appel, These Proceedings.
17. J. Leong, quoted in Ref. 8.
18. In a March 19, 1976 seminar at Fermilab, J. Weiss of the Columbia-Fermilab-Stony Brook Collaboration

showed a clear ψ' peak observed in $p\ Be \to e^+e^-X$ at 400 GeV/c, and inferred $\sigma(\psi')/\sigma(\psi) \sim 10\%$.

19. F. Halzen, Wisconsin preprint C00-501(1976).
20. See, for example, R. Hagedorn, in Atlas of Particle Production Spectra, by H.Grote, R. Hagedorn, and J. Ranft (CERN, Genève, 1970), p. 9.
21. A mountain of evidence appears in Proc. 1975 Int. Symposium on Lepton and Photon Interactions at High Energies, ed. W.T. Kirk (SLAC,Stanford, 1975).
22. This cascade mechanism has been considered by Gaillard, Lee, and Rosner, Ref. 8; implicitly by Einhorn and Ellis, Ref. 9; by F. Gilman, unpublished; and perhaps others. It is discussed at greater length by S.D. Ellis, M.B. Einhorn and C. Quigg, Fermilab-Pub-76/29-THY, and by C.E. Carlson and R. Suaya, William and Mary preprint WM-PP-10(1976).
23. T.W. Appelquist and H.D. Politzer, Phys. Rev. Lett. 34, 43 (1975).
24. T. Appelquist, These Proceedings.
25. S.D. Drell and T.-M. Yan, Phys. Rev. Lett. 25, 316 (1970).
26. C. Friedberg, These Proceedings.
27. P. Mockett, et al., Fermilab proposal P-332; J.Pilcher, et al., Fermilab experiment E-331.
28. R. Blankenbecler, et al., SLAC-PUB-1531 (1975, unpublished). An important misprint is corrected by Minh Duong-Van, Phys. Lett. 60B, 287(1976).
29. G. Altarelli, et al., Nucl. Phys. B92, 413 (1975).
30. G. Sanders, These Proceedings.
31. J.D.Sullivan, These Proceedings.
32. C. Quigg, unpublished calculations.
33. T. Miyazaki and E. Takasugi, Phys. Rev. D8, 2051 (1973); C. Quigg and J.D. Jackson, UCRL-18487, unpublished. The latter contains additional references to earlier estimates.
34. J.D. Bjorken and H. Weisberg, SLAC-PUB-1631; G. Farrar and S.C. Frautschi, CALT-68-518(Rev.).
35. J. Kirz, These Proceedings.
36. R. Johnson, These Proceedings.
37. S. Segler, These Proceedings.
38. L. Leipuner, These Proceedings.
39. R. Mischke, These Proceedings.
40. R. Ruchti, These Proceedings.
41. M. Mallary, These Proceedings.
42. See the talks by S. Segler, T. Yamanouchi, P.A. Piroué, R. Imlay, R. Cence, S. Nurushev, and L.M. Lederman in Proc. XVII Inst. Conf. on High Energy Physics, ed. J.R. Smith, (RHEL, Chilton, 1974), §V. More recent reviews are those by J.W. Cronin, "Review of Direct Lepton Production in Nucleon-Nucleon Collisions," 1975 Erice Lectures, and by L.M. Lederman, in Proc. 1975 Int. Symposium on Lepton and

Photon Interactions at High Energies, ed. W.T. Kirk (SLAC, Stanford, 1975), p. 265.
43. Gaillard, Lee, and Rosner, Ref. 8; A. de Rujula, H. Georgi, and S.L.Glashow, Phys. Rev. D$\underline{12}$, 147 (1975); V.S. Mathur, S. Okubo, and S. Borchardt, Phys. Rev. D$\underline{11}$, 2572 (1975).
44. M. Breidenbach, These Proceedings.
45. K. Pretzl, These Proceedings.
46. H. Rieseberg, These Proceedings.
47. E. Poggio, H.R. Quinn, and S. Weinberg, Phys. Rev. D (to be published).
48. B. Barnett, These Proceedings.
49. M.L. Perl, et al., Phys. Rev. Lett. $\underline{35}$, 1489 (1975).
50. M.B. Einhorn and C. Quigg, Phys. Rev. Lett. $\underline{35}$, 1114 (1975).
51. A.M. Boyarski, et al, Phys. Rev. Lett. $\underline{35}$, 196 (1975).
52. S.L. Glashow, J. Iliopoulos, and L. Maiani, Phys. Rev. D$\underline{2}$, 1285 (1970).
53. J. von Krogh, et al., to be published.
J. Mapp, These Proceedings.
54. H. Deden, et al., Phys. Lett. $\underline{58B}$, 361 (1975); J. Blietschau, et al., Phys. Lett. $\underline{60B}$, 207 (1976).
55. A lengthier discussion may be found in B.W. Lee, "Dimuon Events," in Proc. of the Conf. on Gauge Theory and Modern Field Theory, 1975 (MIT Press, to be published).
56. M.B. Einhorn and B.W. Lee, Fermilab-Pub-75/56; M.K. Gaillard, S. Jackson, and D. Nanopoulos, CERN TH. 2049; V. Barger and T. Weiler, Wisconsin preprint COO-456.
57. See Gaillard, Lee, and Rosner, Ref. 8 for a more complete discussion.
58. E. Eichten, et al., Phys. Rev. Lett. $\underline{36}$, 500 (1976).
59. B. Wiik, in Proc. 1975 Int. Symposium on Lepton and Photon Interactions at High Energies, ed. W.T. Kirk (SLAC, Stanford, 1975), p. 69. J.Heintze, ibid., p. 97.
60. D.C. Hom, et al., Fermilab-Pub-76/19-EXP.

LIST OF PARTICIPANTS

C. Akerlof, University of Michigan
E. W. Anderson, Iowa State University
J. Appel, Fermilab
T. Appelquist, Yale University
N. Baggett, Purdue University
B. Barish, California Institute of Technology
B. Barnett, Johns Hopkins/University of Maryland
N. N. Biswas, Notre Dame
B. B. Brabson, Indiana University
M. Breidenbach, SLAC
A. Benvenuti, University of Wisconsin
T. Bowen, University of Arizona
R. Burnstein, Illinois Institute of Technology
E. G. Cazzoli, BNL
M. Chen, M.I.T.
W. Chinowsky, LBL
H. O. Cohn, Oak Ridge National Laboratory
J. Cronin, University of Chicago
D. Cutts, Brown University
S. R. Deans, South Florida University
R. Dulude, Brown University
S. Ellis, University of Washington
R. Endorf, University of Cincinnati
T. Ferbel, University of Rochester
H. Fenker, Vanderbilt University
C. Friedberg, LBL
K. J. Foley, BNL
T. Fieguth, SLAC
I. Gaines, Fermilab
T. K. Gaisser, BNL
D. Geffen, University of Minnesota
M. Goldberg, Syracuse University
S. Hagopian, Florida State University
J. Hanlon, Stony Brook
T. Handler, University of Tennessee
R. Harris, Fermilab
E. L. Hart, University of Tennessee
E. Harvey, University of Wisconsin
J. Hauptman, UCLA
D. Hood, Purdue University
V. Highland, Temple University
W. Holladay, Vanderbilt University
R. Hulsizer, M.I.T.
F. R. Huson, Fermilab
K. Johnson, M.I.T.
K. Johnson, Stony Brook
R. Johnson, Fermilab
S. Kahn, Fermilab

J. Kirz, Stony Brook
A. Kreymer, Indiana University
J. E. Lannuti, Florida State University
L. B. Leipuner, BNL
S. Longo, La Salle College
M. Longo, University of Michigan
H. Lubatti, University of Washington
M. H. MacGregor, Lawrence Livermore Lab
M. Mallary, Northeastern University
J. E. Mandula, M.I.T.
J. Mapp, University of Wisconsin
M. Marx, BNL
J. Marraffino, Vanderbilt University
J. Matthews, Michigan State University
G. C. Moneti, Syracuse University
N. A. McCubbin, Rutherford Lab
B. Meadows, University of Cincinnati
R. E. Mickens, Fisk University
R. E. Mischke, Los Alamos
M. Misbina, Yale University
H. Paik, Indiana University
C. Pang, University of Illinois
R. Panvini, Vanderbilt University
D. L. Parker, Iowa State University
J. Patrick, Vanderbilt University
W. T. Pinkston, Vanderbilt University
S. Poucher, Vanderbilt University
K. Pretzl, Max Planck Institute
S. D. Protopopescu, BNL
C. Quigg, Fermilab
S. Reucroft, Vanderbilt University
H. Rieseberg, Heidelberg University
D. Ritson, SLAC
A. Rogers, Vanderbilt University
C. Rubbia, Harvard University
R. Ruchti, Northwestern University
E. O. Salant, BNL
G. Sanders, Princeton University
J. Scanio, University of Cincinnati
W. Scott, Fermilab
S. L. Segler, Rockefeller University
K. C. Stanfield, Purdue University
E. Smith, Vanderbilt University
G. Snow, University of Maryland
C. Sorenson, ANL
P. H. Steinberg, University of Maryland
S. Stone, Vanderbilt University
J. D. Sullivan, University of Illinois
P. Suranyi, University of Cincinnati
W. W. Wada, Ohio State

K-H Wang, McGill University
Y. Watanabe, ANL
M. S. Webster, Vanderbilt University
L. Witten, University of Cincinnati
M. Weinstein, SLAC
H. Weisberg, University of Pennsylvania
S. White, Rockefeller University

AIP Conference Proceedings

		L.C. Number	ISBN
No. 1	Feedback and Dynamic Control of Plasmas (Princeton) 1970	70-141596	0-88318-100-2
No. 2	Particles and Fields - 1971 (Rochester)	71-184662	0-88318-101-0
No. 3	Thermal Expansion - 1971 (Corning)	72-76970	0-88318-102-9
No. 4	Superconductivity in d- and f-Band Metals (Rochester 1971)	74-18879	0-88318-103-7
No. 5	Magnetism and Magnetic Materials - 1971 (2 parts) (Chicago)	59-2468	0-88318-104-5
No. 6	Particle Physics (Irvine 1971)	72-81239	0-88318-105-3
No. 7	Exploring the History of Nuclear Physics (Brookline, 1967, 1969)	72-81883	0-88318-106-1
No. 8	Experimental Meson Spectroscopy - 1972 (Philadelphia)	72-88226	0-88318-107-X
No. 9	Cyclotrons - 1972 (Vancouver)	72-92798	0-88318-108-8
No.10	Magnetism and Magnetic Materials - 1972 (2 parts) (Denver)	72-623469	0-88318-109-6
No.11	Transport Phenomena - 1973 (Brown University Conference)	73-80682	0-88318-110-X
No.12	Experiments on High Energy Particle Collisions - 1973 (Vanderbilt Conference)	73-81705	0-88318-111-8
No.13	π-π Scattering - 1973 (Tallahassee Conference)	73-81704	0-88318-112-6
No.14	Particles and Fields - 1973 (APS/DPF Berkeley)	73-91923	0-88318-113-4
No.15	High Energy Collisions - 1973 (Stony Brook)	73-92324	0-88318-114-2
No.16	Causality and Physical Theories (Wayne State University, 1973)	73-93420	0-88318-115-0
No.17	Thermal Expansion - 1973 (Lake of the Ozarks)	73-94415	0-88318-116-9
No.18	Magnetism and Magnetic Materials - 1973 (2 parts) (Boston)	59-2468	0-88318-117-7
No.19	Physics and the Energy Problem - 1974 (APS Chicago)	73-94416	0-88318-118-5
No.20	Tetrahedrally Bonded Amorphous Semiconductors (Yorktown Heights, 1974)	74-80145	0-88318-119-3
No.21	Experimental Meson Spectroscopy - 1974 (Boston)	74-82628	0-88318-120-7
No.22	Neutrinos - 1974 (Philadelphia)	74-82413	0-88318-121-5
No.23	Particles and Fields - 1974 (APS/DPF Williamsburg)	74-27575	0-88318-122-3
No.24	Magnetism and Magnetic Materials - 1974 (20th Annual Conference San Francisco)	75-2647	0-88318-123-1
No.25	Efficient Use of Energy (The APS Studies on the Technical Aspects of the More Efficient Use of Energy)	75-18227	0-88318-124-X
No.26	High-Energy Physics and Nuclear Structure - 1975 (Santa Fe and Los Alamos)	75-26411	0-88318-125-8
No. 27	Topics in Statistical Mechanics and Biophysics: A Memorial to Julius L. Jackson (Wayne State University-1975)	75-36309	0-88318-126-6
No. 28	Physics and Our World: A Symposium in Honor of Victor F. Weisskopf (M.I.T. 1974)	76-7207	0-88318-127-4
No. 29	Magnetism and Magnetic Materials - 1975 (21st Annual Conference, Philadelphia)	76-10931	0-88318-128-2
No. 30	Particle Searches and Discoveries - 1976 (Vanderbilt Conference)	76-19949	0-88318-129-0
No. 31	Structure and Excitations of Amorphous Solids (Williamsburg, Va., 1976)		0-88318-130-4